DEEP SPACE

**Beyond the Solar System
to the Edge of the Universe
and the Beginning of Time**

BLACK DOG
& LEVENTHAL
PUBLISHERS
NEW YORK

Govert Schilling

Published by
Black Dog & Leventhal Publishers, Inc.
151 West 19th Street
New York, NY 10011

Distributed by
Workman Publishing Company
225 Varick Street
New York, NY 10014

Manufactured in China

Cover and interior concept and design by Matthew Riley Cokeley

Cover and interior design and layout by Sheila Hart Design

Cover photograph courtesy ESA / Hubble / NASA

ISBN-13: 978-1-57912-978-1

h g f e d c b a

Library of Congress Cataloging-in-Publication Data

Contents

cosmic ocean. It is our tiny backyard, from which we can gaze out into the cosmic equivalents of our street, our town, our country, and our world.

Marcus Chown's book, *Solar System*, published by Black Dog & Leventhal in 2011, took readers on a visually stunning exploratory tour of Earth's siblings and their retinue of satellites. This book, after briefly looking around in the planetary back-yard, takes you much farther, into the realm of stars and nebulae, pulsars and stellar clusters, supernova explosions and black holes, galaxies and clusters, all the way to the edge of space and the beginning of time. Wil Tirion's Star Atlas at the back of this book will help you find your way across the night sky.

Working our way out from the here and now to the distant there and then, we encounter numerous famous inhabitants of the Universe, like the star Betelgeuse, the Orion Nebula, the Pleiades cluster, and the Andromeda galaxy. But this is not just a cosmic "Facebook." Astronomers now understand how all these objects are mutu-ally related. Together, they tell one exciting story of cosmic evolution, from the first density fluctuations in the Big Bang; through the formation of galaxies; to the birth of stars, habitable planets, and, eventually, life.

With the discovery of weird objects like blue stragglers, magnetars, and quasars, and with mystery constituents like dark matter and dark energy, the diary of the Universe has become ever more intricate. Most of the contents of this book could not have been written 25 years ago, simply because of our limited knowledge back then.

And even if it could have been written, you would miss out on the gorgeous photographs that have been provided by large observatories on the ground and telescopes in space. We are fortunate to live at a time when the sensitive eyes of the Hubble Space Telescope and the European Very Large Telescope in Chile present us with a wealth of images that fire our imagination as much as a Rembrandt or Van Gogh might.

It has been a privilege to work on this book, building on Marcus Chown's earlier volume about the solar system. In writing the text and selecting the images, I real-ized once again what a miraculous Universe we find ourselves in. Join me on a voyage through Deep Space, enjoy the views, and learn about the wide and wondrous world you live in.

◄ Resembling a budding rose on a stem—a true deep space flower— these two galaxies, at a distance of 300 million light-years from Earth, are distorted by their mutual gravity. The smaller one may have passed right through the larger one a few hundred million years ago.

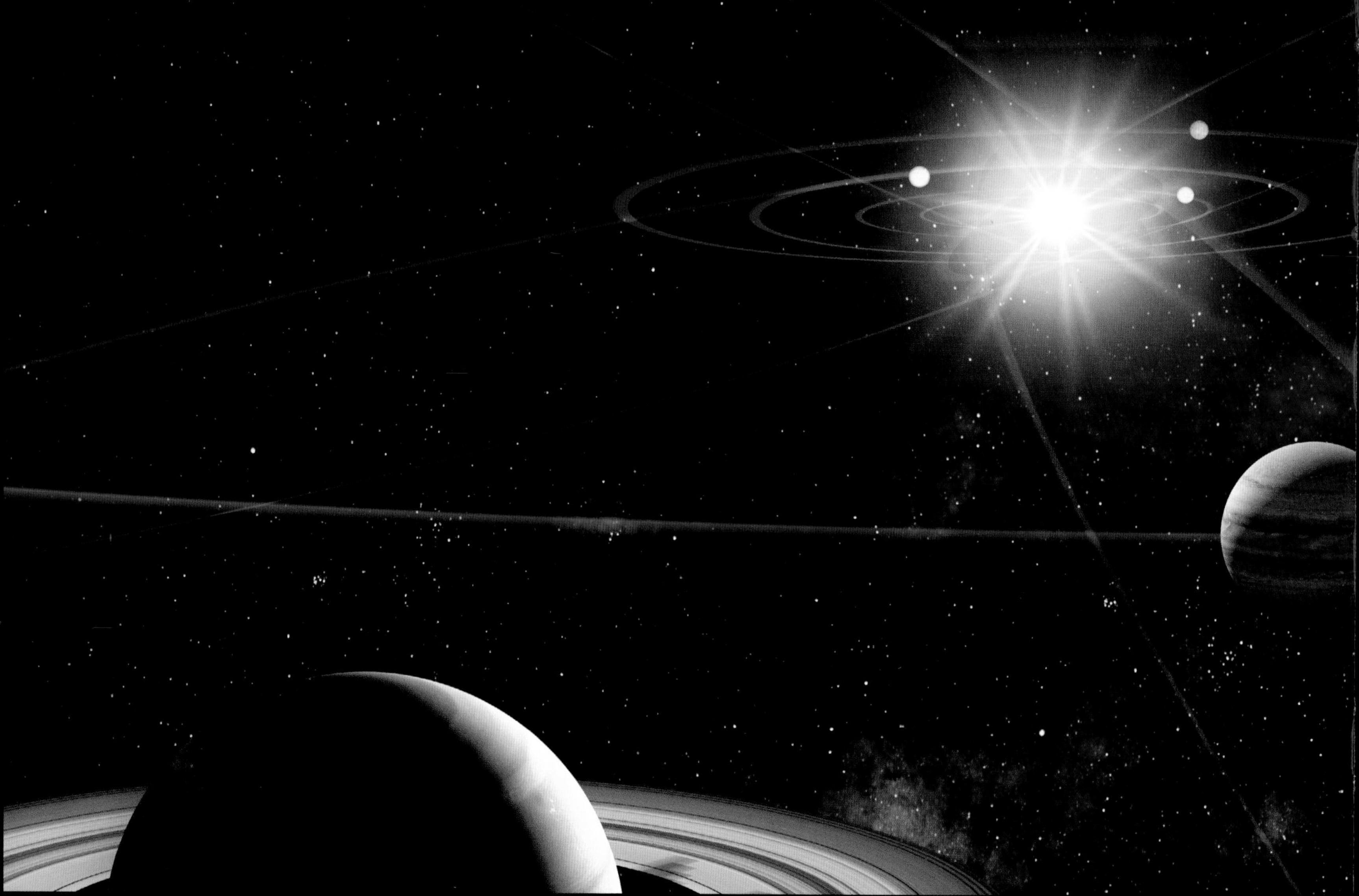

The SOLAR SYSTEM

It's all about the Sun—both literally and figuratively. The Sun accounts for 99% of all the mass in the solar system and consists almost entirely of hydrogen and helium, the two lightest elements known to nature. The remaining 1%—planets, moons, asteroids, ice dwarfs, and comets—is nothing more than the residue of the birth of a single star. Our own Earth is like a grain of sand on the flanks of a volcano.

The solar system is our cosmic backyard; the Sun is our mother, and the planets are our brothers and sisters. It is a well-organized family, with four small, rocky planets close to the Sun and four giant gas planets much farther away. All the planets orbit the Sun in the same direction and almost in the same plane. It is only since we discovered other planetary families in the Milky Way that we have come to realize that the tranquility, purity, and regularity of our own living environment is by no means something we should take for granted.

One planet in the solar system stands out. That is, of course, Earth, with its oceans of liquid water, its oxygen-rich atmosphere, and its teeming tapestry of life-forms, from microscopic single-celled organisms to giant sequoias and blue whales. Our existence is closely intertwined with the life of the Universe. Our body cells are made up of cosmic material; the building blocks of life were brought to Earth by comets and meteorites. At the same time, we are constantly under threat from cosmic disasters that could just as easily destroy all life on the "blue planet."

◄ In a remote corner of our Milky Way galaxy, our solar system is home to a wide variety of planets, including Earth.

PASSPORT

Name: Sun

Diameter:
1,392,000 km

Rotation period:
25.4 days

Mass: 328,946 x Earth

Surface gravity:
27.9 x Earth

Age: 4.6 billion years

Distance:
149.6 million km

Solar Puzzles

The Universe has some ten quintillion stars. Less than a hundred-millionth of a percent of those are to be found in the Milky Way, but that still amounts to a few hundred billion. One of those small stars in the Milky Way is our Sun, the source of energy for all life on Earth. It is nothing more than a "pinprick" in the dark expanse of the cosmos, but it is indispensable in bringing us light and heat.

Stars like the Sun are surprisingly simple in their makeup. They consist of roughly 75% hydrogen and 24% helium. Only 1% of the Sun comprises heavier atoms. All that gas is compressed into a sphere as a result of its own gravity. Temperature and pressure increase as you penetrate deeper into that ball, and conditions are so extreme in the center that spontaneous nuclear reactions occur. The energy that is released radiates from the incandescent surface of the Sun in the form of light and heat. It's as simple as that.

Yet the Sun still presents us with many puzzles. No one knows exactly how the Sun's rarified atmosphere, the corona, can be heated to more than a million degrees. Distorted magnetic fields cause both cool, dark sunspots and powerful solar flares that blow high-energy, electrically charged particles into space; yet why the Sun's eleven-year activity cycle sometimes skips a few beats and how the Sun affects the Earth's climate remain unclear. One thing we do know, however, is that an extremely powerful solar flare could fry power grids, collapse communication networks, and completely disrupt our vulnerable technological society.

▲ Between April 2012 and April 2013, active regions of the Sun preferentially occurred north and south of the equator, as evidenced by this composite of twenty-five extreme ultraviolet images.

▶ Gas temperatures are color-coded in this multi-wavelength view of the Sun, obtained on March 30, 2010: red is relatively cool; blue and green are hot.

▼ A huge ribbon of sizzling gas spews away from an active area on the Sun's surface.

▶ The Swedish Solar Telescope at La Palma, Canary Islands, Spain, captures fine details in the outer, penumbral regions of sunspots.

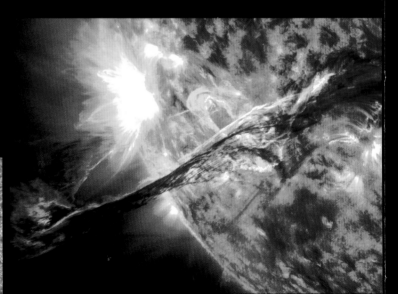

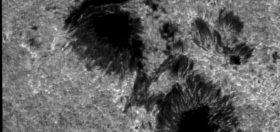

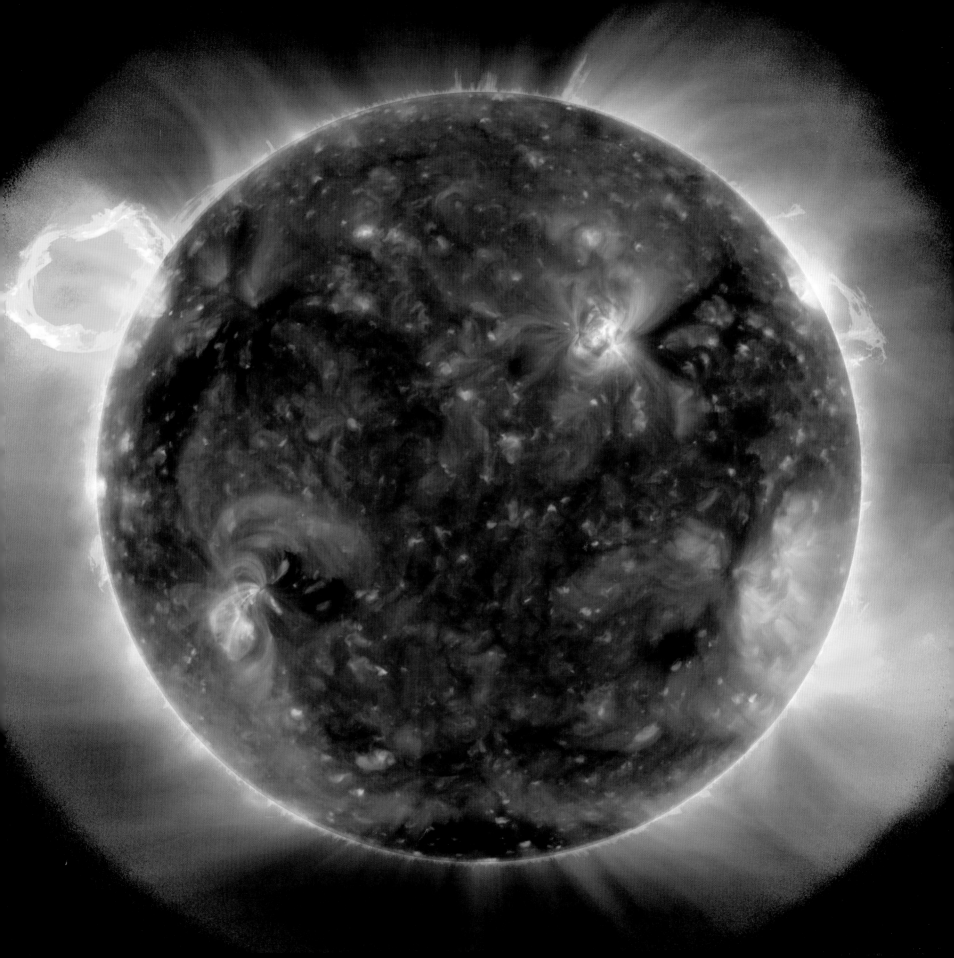

Name: Mercury

Distance to Sun:
57.9 million km

Orbital period:
88 days

Diameter: 4,880 km

Rotation period:
$58^d\ 15^h\ 31^m$

Mass: 0.055 x Earth

Surface gravity:
0.37 x Earth

Number of moons: 0

Name: Venus

Distance to Sun:
108.2 million km

Orbital period:
225 days

Diameter: 12,103 km

Rotation period:
$243^d\ 00^h\ 27^m$

Mass: 0.815 x Earth

Surface gravity:
0.91 x Earth

Number of moons: 0

Name: Earth

Distance to Sun:
149.6 million km

Orbital period:
1 year

Diameter: 12,756 km

Rotation period:
$23^h\ 56^m\ 04^s$

Mass: 6×10^{24} kg

Surface gravity:
1 g

Number of moons: 1

Name: Mars

Distance to Sun:
228 million km

Orbital period:
1.88 years

Diameter: 6,794 km

Rotation period:
$24^h\ 37^m\ 23^s$

Mass: 0.11 x Earth

Surface gravity:
0.38 x Earth

Number of moons: 2

Marbles of Iron and Rock

The four inner planets of our solar system—Mercury, Venus, Earth, and Mars—are also known as the *terrestrial planets*. On the outside they all look quite different, but on the inside they are similar, with a metal core of iron and nickel and a mantle of rock. They are all made up of heavy elements, and that is exactly why they are so small. The interstellar cloud of gas and dust from which the Sun and the planets were created contained few heavy elements. In the inner regions of the solar system, where the radiation of the newborn Sun blew the volatile gases into space, little heavy material was left to form planets.

Each of the four inner planets bears the marks of primordial cosmic violence, in the form of impact craters, even though on Earth these scars have largely been erased by erosion and geological activity. A large part of Mercury's rocky mantle was probably destroyed during a catastrophic collision sometime in the distant past, leaving the planet with a relatively large iron core. Venus' backward rotation, the origin of our Moon, and the disappearance of Mars' formerly denser atmosphere may also have been caused by similar cosmic disasters. Although Mercury has probably always been a bone-dry, smooth marble, Venus, Earth, and Mars were much more similar to each other more than four billion years ago. But Venus "boiled dry" as the result of a runaway greenhouse effect, and Mars lost its oceans and cooled down to become a frozen desert of rock. Life could sustain itself only on Earth.

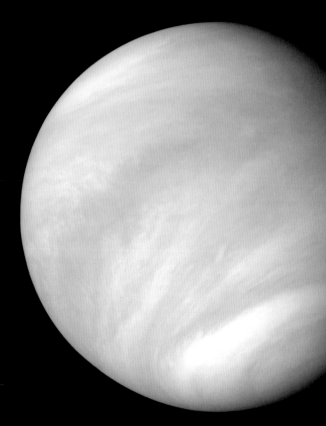

▼ The surface of Venus is hidden from view by a dense carbon dioxide atmosphere with thick clouds of sulfuric acid.

▼ NASA's Mars rover Spirit was descending from "Husband Hill" when it shot this detailed panorama of the red planet in November 2005.

◄ The terrestrial planets are shown here to scale. Mercury, with its huge core of nickel and iron, is the smallest of the four.

► Mars may be the planet most like Earth, but it lacks liquid surface water and a substantial atmosphere.

▲ Colors represent subtle variations in altitude in this volcanic plain on Mercury's northern hemisphere.

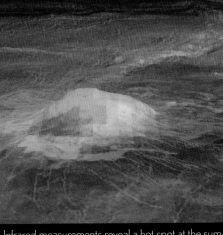

◄ Warm and wet, the Blue Marble, as it is called, is the only planet in the solar system that harbors abundant life.

▲ Infrared measurements reveal a hot spot at the summit of a shield volcano on Venus. The planet might still be geologically active.

PASSPORTS

Name: Jupiter

Distance to Sun:
778.2 million km

Orbital period:
11.86 years

Diameter: 142,200 km

Rotation period:
9ʰ 55ᵐ 30ˢ

Mass: 317.8 x Earth

Surface gravity:
2.37 x Earth

Number of moons: 65

Name: Saturn

Distance to Sun:
1.43 billion km

Orbital period:
29.46 years

Diameter: 120,500 km

Rotation period:
10ʰ 39ᵐ 22ˢ

Mass: 95.2 x Earth

Surface gravity:
0.93 x Earth

Number of moons: 62

Name: Uranus

Distance to Sun:
2.86 billion km

Orbital period:
84.02 years

Diameter: 51,120 km

Rotation period:
17ʰ 14ᵐ 24ˢ

Mass: 14.5 x Earth

Surface gravity:
0.89 x Earth

Number of moons: 27

Name: Neptune

Distance to Sun:
4.48 billion km

Orbital period:
164.77 years

Diameter: 49,530 km

Rotation period:
15ʰ 57ᵐ 59ˢ

Mass: 17.1 x Earth

Surface gravity:
1.12 x Earth

Number of moons: 14

Giants of Ice and Gas

In the colder outer regions of the solar system, much more material was available to build planets. Volatile molecules like water vapor, methane, and ammonia condensed here to ice crystals that were not as easily blown away by the radiation from the newborn Sun. This led to the formation of the relatively large and heavy cores of the four giant planets. Their gravitational pull attracted great quantities of hydrogen and helium gas. The outcome of this process was four colossal giant planets of gas without a solid surface and with enormous storms raging in their thick atmospheres.

The Great Red Spot in Jupiter's atmosphere, the mysterious hexagon hurricane around Saturn's north pole, the storm systems, and unimaginable wind speeds (exceeding 2,000 kilometers per hour) on Neptune are all natural phenomena compared with which the Earth's cyclones and tornadoes are storms in a teacup. Conditions inside these giant planets are also unearthly: the gases in the deepest interiors of Jupiter and Saturn have been compressed together with such force that they have become liquid and have even developed metal-like properties; the mantles of Uranus and Neptune consist of supercompact "warm ice" of water, ammonia, and methane.

Jupiter and Saturn have been studied up close over a long period by the unmanned American space probes Galileo and Cassini. We know much less about Uranus and Neptune, which were not discovered until 1781 and 1846, respectively. In 1986 and 1989, Voyager 2 paid a fleeting visit to these two ice giants, but all other research into the two outermost planets in the solar system has been conducted with telescopes on or in orbit around the Earth.

◄ Giant Jupiter, with its impressive clouds and storm systems, is by far the largest and most massive planet in the solar system.

▶ Remote Neptune, seen here by Voyager 2 during its flyby in August 1989, is the outermost planet in the solar system.

▶ Jupiter's Great Red Spot, wide enough to swallow two Earths at once, is a giant anticyclonic storm that has been raging for centuries.

◀ To Voyager 2, the spacecraft that flew past the planet in January 1986, Uranus was little more than a bland atmospheric world.

▲ The swirling hurricane at Saturn's north pole, captured here in near-infrared light, measures 2,000 kilometers across.

◀ During Saturn's equinox, the planet and its rings are illuminated exactly from the

> Enceladus is a small, icy moon of Saturn. It is only some 500 kilometers across, but it harbors subsurface seas that might host microbial life.

▲ Saturn's moon Hyperion is only a few hundred kilometers across and appears to be an irregularly shaped, porous mass of frozen water.

▼ Cracks and ridges in the icy surface of Jupiter's moon Europa, here shown in false color, reveal the existence of a subsurface ocean of liquid water.

Deep Space

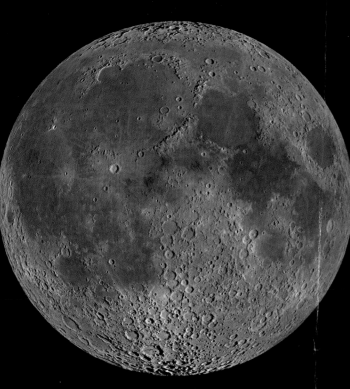

Vassals for the Planets

Galileo Galilei discovered in 1610 that our Moon, which orbits Earth, is not unique. Other planets have moons, too; Galileo discovered four around the giant planet Jupiter. Thanks to more powerful telescopes and more sensitive cameras, increasing numbers of planetary satellites have since been found, and many small ones have been identified by the Voyager probes. The current count (November 2013) is 180, nineteen of which are larger than 500 kilometers in diameter.

Mercury and Venus have no natural satellites. Our own Moon was probably formed from the fragments of a catastrophic cosmic collision, and the two small moons orbiting Mars are almost certainly captured asteroids. The many small satellites circling the giant planets, which often move in very irregular orbits, are likely captured objects, as is the colossal Neptunian moon Triton. But the large moons of Jupiter, Saturn, and Uranus, with their regular orbits, were formed at the same time as their host planets, as if they are part of their own mini-solar system.

The many dozens of planetary satellites found in the solar system have an astounding diversity of geological properties, from the active sulfur volcanoes of Jupiter's moon Io to the ice fountains of Saturn's Enceladus and the steep ice cliffs of Miranda, which orbits Uranus. By far the most interesting, however, is Saturn's moon, Titan, the only planetary moon with a substantial atmosphere and lakes of liquid methane. Many asteroids and ice dwarfs are also accompanied by moons of varying sizes; that certainly says something about how they were formed, but exactly what that is we do not yet know.

▲ Our moon bears the scars of cosmic impacts, most of which happened hundreds of millions of years ago.

▲ Infrared observations and radar measurements have revealed the cold surface of Saturn's giant moon Titan, which holds lakes of liquid methane.

▼ Phobos, the larger of the two small moons of Mars, is a potato-shaped rock, battered by meteorites.

◄ Voyager scientists at NASA's Jet Propulsion Laboratory nicknamed Jupiter's moon Io the "pizza moon." No other world in the solar system has the same level of volcanic activity.

Running Rings around the Giants

▲ Infrared observations with the 10-meter Keck telescope at Mauna Kea, Hawaii, reveal the thin, dark rings of Uranus.

▲ Neptune's ring arcs were captured in August 1989 by NASA's Voyager 2.

The giant planets are not only accompanied by a large number of satellites, but each of them also has a system of rings consisting of dust, rubble, and lumps of ice. Only the rings of Saturn can be seen from Earth with a small telescope. They were observed by Galileo in Italy in 1610, but Dutch astronomer Christiaan Huygens, in the mid-seventeenth century, was the first to understand their true nature.

The dust particles in the tenuous ring of Jupiter are produced by impacts of micrometeorites on the planet's four inner moons. The origin and lifespan of the dark dust rings and incomplete ring arcs orbiting Uranus and Neptune are less clear. The broad, bright rings of Saturn consist of both small and large lumps of rock and ice that are possibly the remains of a pulverized small satellite.

Since 2004, the space probe Cassini has been studying the rings of Saturn closely, which has not only produced magnificent photos but has provided fascinating insights into the interaction between the satellites and the particles in the rings, the origin of density waves and undulations, and the ease with which small particles can clump together to form large structures. Saturn's ring system is a unique laboratory for studying processes in other disklike structures, such as spiral galaxies and protoplanetary disks around young stars.

In around ten million years, the planet Mars will also be surrounded by a ring of rubble: Mars' moon, Phobos, is spiraling slowly but surely inward and, in the distant future, will be torn to pieces by tidal forces.

▼ Saturn's impressive ring system may look like an old-school phonograph record, but it is much thinner in comparison: the rings have a diameter of 275,000 kilometers but a thickness of a mere 20 meters.

➤ Backlit by the Sun, the tenuous dust rings of Jupiter—and the planet's atmosphere—are visible in this Voyager image.

▼ NASA's Cassini spacecraft flew through Saturn's shadow when it shot this image of the planet's night side and backlit ring system. Just above the left edge of the main rings is Earth.

➤ Gravitational perturbations by small, embedded moons can lift ring particles from the central plane, creating a stunning shadow play.

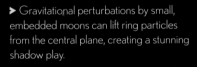

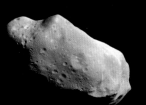

> Asteroids Ida, Lutetia, Itokawa, and Gaspra and comets Hartley 2 and Tempel 1 are some of the solar system's small bodies that have been studied in close-up. These photographs are not to scale.

Small is Beautiful

On the first day of the nineteenth century, Giuseppe Piazzi discovered a new "planet" between the orbits of Mars and Jupiter. Ceres, as the new celestial body was called, was soon joined by Pallas, Juno, and Vesta. It was only half a century later, when astronomers discovered many more small objects in similar orbits, that they realized that the four "planets" did not qualify for that status, and they were classified as *asteroids*.

In 2006, history repeated itself when Pluto, which had been known as the ninth planet since its discovery in 1930, lost its planetary status after a large number of other icy objects were discovered beyond the orbit of Neptune; these objects had similar sizes and properties and in comparable eccentric and inclined orbits.

The rocky objects in the asteroid belt beyond the orbit of Mars are remnants of the formation of the terrestrial planets. The ice dwarfs in the Kuiper Belt beyond the orbit of Neptune are the remains of the formation of the giant planets. As a result of gravitational disturbances, countless lumps of ice— the comets—were catapulted out of the solar system and accumulated in the spherical Oort Cloud, which extends almost halfway to the nearest star.

The orbits of the asteroids can also be disturbed so that they end up in the inner regions of the solar system and can collide with the Earth. Since the first visits to the nucleus of a comet (in 1986, Halley's comet) and the first flyby of an asteroid (in 1991, Gaspra), unmanned space research into these small celestial bodies has really taken off and offers us a glimpse of the time when the solar system was still being formed.

◄ Nicknamed "the Snowman," this series of three impact craters on the surface of asteroid Vesta was shot in false color—to reveal variations in surface composition—by the Dawn spacecraft.

Deep Space

▲ Sight lines in neolithic Stonehenge mark the points on the horizon where the Sun and the Moon rise and set at equinoxes and solstices.

▼ The great Egyptian pyramids exactly face north, east, south, and west; some slanting tunnels appear to be oriented at the stars.

◄ The launch of Sputnik 1, in 1957, ushered in a whole new era in the history of astronomy.

► This portrait of Galileo Galilei was painted by Justus Sustermans 26 years after the Italian astronomer first trained a telescope on the sky.

The History of Astronomy

Astronomy is as old as humanity itself. Tens of thousands of years ago, our distant forebears must have gazed in wonder at the glittering night sky and the regular cycle of day and night, summer and winter. The cosmos (*cosmos* is the Greek word for "order") was a place of imperishable, divine perfection.

A few thousand years ago in Babylon, between the two great rivers Euphrates and Tigris (in present-day Iraq), the first steps were taken to keep a systematic record of the motion of the Sun, the Moon, and the planets. Anyone who understood the acts of the gods, the ancient Babylonians reasoned, could also know what was going to happen here on Earth.

Today's astronomers no longer believe in astrology, but it was these first Babylonian astrologers who laid the basis for what would later become the science of astronomy. Much of their knowledge was taken over by the Greeks who, 2,500 years ago, resolved to develop a view of the world in which cosmic movements played an integral role.

The great pioneer of this Greek worldview was Claudius Ptolemy, who set out his ideas in the Almagest. According to Ptolemy, Earth sat motionless at the center of the Universe, and the Sun, the Moon, and the planets circled around it in complicated orbits.

It was not until the mid-sixteenth century that Polish astronomer Nicolaus Copernicus came up with an alternative view of the Universe, with the Sun (helios) at the center. According to Copernicus, Earth is just one of the planets orbiting the Sun. A century later, this heliocentric worldview has been widely accepted, although the church found it difficult to accept a theory that placed the "sinful" Earth on a par with the "divine" planets.

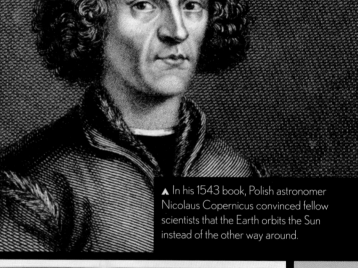

▲ In his 1543 book, Polish astronomer Nicolaus Copernicus convinced fellow scientists that the Earth orbits the Sun instead of the other way around.

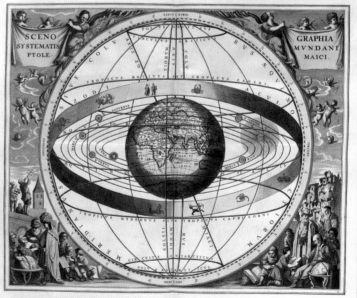

➤ A seventeenth-century engraving depicts the then-abandoned Earth-centered view of the Universe.

Half a century after Copernicus' death, the telescope was invented in the Netherlands. Italian astronomer Galileo Galilei discovered millions of new stars in the Milky Way, mountains on the Moon, spots on the Sun, moons around Jupiter, and the phases of Venus. Johannes Kepler established the mathematical rules of planetary motion; Isaac Newton brought order to the solar system with his universal theory of gravity; and astronomers discovered comets, nebulae, and even a number of new planets.

Improved observation techniques enabled astronomers to determine the distance to the stars. The invention of the spectroscope—an instrument to dissect and study the light of a star—allowed them to learn more about the composition of celestial bodies. New insights in physics and chemistry in the nineteenth century also proved applicable in astronomy.

For a little less than a century, we have known that the Universe is enormously larger than our own Milky Way and that it is still expanding: the distances between galaxies are increasing over time. Present-day astronomers no longer try to discover how planets move or how far away the stars are. Instead, they discuss black holes, life on other planets, dark matter, and the origin of the Universe.

Our fascination with the Universe has not diminished over the centuries. On the contrary, the world of the cosmos seems to become ever more fascinating with each new discovery.

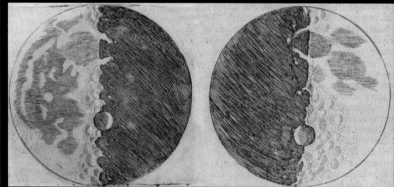

⌃ Using a simple homemade telescope, Galileo was the first to make detailed drawings of craters and mountains on the moon.

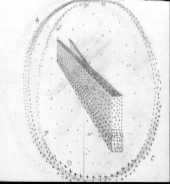

➤ In the eighteenth century, William Herschel used star counts to derive a (faulty) three-dimensional model of our Milky Way galaxy.

NASA's Curiosity rover, which is studying the past habitability of Mars, is an icon of twenty-first-century space science.

The History of Astronomy

The
BIRTH
of Stars

A few hundred years ago, it was unimaginable to think that stars do not live forever but instead were once "born" and will one day "die." Today we know that nothing in the Universe lasts forever. The night sky looks very different today from how it looked a billion years ago or will look a billion years in the future. Where we now see cold, dark clouds of gas and dust, new stars will one day twinkle.

A star like the Sun has a life-span of some ten billion years. So it is not surprising that the birth of a star also takes longer than that of a human baby. After the first contractions have started, it can take tens of thousands of years for the new stellar infant to utter its first cry. Although little seems to happen in the cosmic nursery during all that time, appearances can be deceptive: our lives are simply too short to witness it.

These cosmic nurseries are among the most photogenic objects in the Universe. As the first stars are being born, the surrounding gas is illuminated in all colors of the rainbow, and irregularly shaped dust clouds stand out against a background of glowing nebular filaments. These spectacular displays also give us an astounding peep at the Sun's baby album: more than four and a half billion years ago, our own star—probably together with dozens of brothers and sisters—was created in a similar way from a cosmic cloud of gas and dust.

◄ Named after its spidery appearance, the Tarantula Nebula in the Large Magellanic Cloud is one of the grandest star-forming regions in the local Universe.

A Cold, Dark Nursery

▲ At submillimeter wavelengths, the European Atacama Pathfinder Experiment (APEX) telescope in Chile reveals the faint glow of dark dust clouds in the Orion Complex.

The space between the stars is not empty, although it contains less gas and dust than the best vacuum that can be created in laboratories on Earth. Sometimes, gravity sweeps interstellar material together to form gigantic clouds. In the heart of these clouds, where stellar light can hardly penetrate, it is pitch dark and deathly cold. In such conditions, individual atoms can combine to form simple molecules. These extended structures are therefore known as *molecular clouds*.

The closest molecular cloud to Earth, around 1,500 light-years away, is the Orion Complex. This colossal dark cloud is a few hundred light-years across and, in the night sky, covers practically the entire constellation Orion. Most of the complex is, of course, invisible with a normal telescope, but it does contain glowing gas nebulae and newborn stars; some dark clouds are visible in silhouette against the bright nebular background.

The cold dust in the cloud, however, emits radiation at longer wavelengths, which can be imaged using a submillimeter telescope, like the Atacama Pathfinder Experiment (APEX) telescope in northern Chile, which is located at an altitude of 5,000 meters. When the APEX readings (orange in the photo) are superposed on a "regular" image, it can clearly be seen that the submillimeter radiation comes from the darkest areas of the Orion Complex.

Just as an X-ray shows us our skeleton, the submillimeter photo shows us what is happening in the deepest interior of the Orion Complex. The giant molecular cloud contains sufficient gas and dust to form many thousands of stars.

◄ The well-known constellation of Orion is filled with dark dust and faintly glowing gas, as seen in this long-exposure photograph.

PASSPORT

Name: Rho Ophiuchi Cloud Complex

Constellation: Ophiuchus

Sky position:
R.A. 16ʰ 28ᵐ 06ˢ
Dec.-24° 32.5'

Star chart: 12

Distance: 450 light-years

Star Formation in our Backyard

A little to the north of the bright star Antares in the constellation Scorpio is an irregularly formed complex of gas and dust clouds containing dozens of newborn stars. The complex, called the Rho Ophiuchi Cloud after the nearby star in the constellation Ophiuchus, is one of the closest star-forming regions to Earth, at a distance of about 450 light-years. It contains enough material to form a few thousand stars like the Sun. The cloud, with its fantastic range of colors, is a favorite of astrophotographers.

Photos taken in visible light show wispy dust clouds that completely absorb the light of more distant stars. The high density of these filaments was probably caused by shock waves from a neighboring star-forming region. When such shock waves propagate through interstellar space, they can cause local density enhancements, which can lead to the birth of new stars.

The American Wide-field Infrared Survey Explorer (WISE) space telescope made infrared photos of the Rho Ophiuchi cloud. The colors on the infrared images do not correspond to those on the optical photograph; the red spot at the bottom right is a reflection nebula around the star Sigma Scorpii, which can be seen as a blue-white star on the right edge of the optical image.

Most of the dust in the Rho Ophiuchi cloud is practically transparent to WISE's infrared eyes, making it possible to see a number of newborn stars left of center. In the infrared image, these Young Stellar Objects (YSOs) are seen as pink stars. In visible light, they are completely concealed by the dust clouds surrounding them.

▲ At optical wavelengths, the Rho Ophiuchi star-forming region is a chaotic mix of hot gas and dark tendrils of cold dust.

029

The Birth of Stars

PASSPORT

||||||||||||||||||||||||||||||||||||||||

Name: Orion Nebula
M42, NGC 1976

Constellation: Orion

Sky position:
R.A. 05ʰ 35ᵐ 17ˢ
Dec. -05° 23.5'

Star chart: 9

Distance: 1,350 light-years

Diameter: 25 light-years

Stellar Forest Fire in Orion

At the end of the nineteenth century, astronomers thought they had encountered a new chemical element. The pale-green color of certain parts of the Orion Nebula could not be explained by the chemistry of the time. The new element was called *nebulium*. It was not until 30 years later that the green light was found to be produced by oxygen atoms that had been doubly ionized, which requires a combination of extremely low density and very high temperature; certain parts of the rarefied Orion Nebula have temperatures of more than 10,000°C.

The Orion Nebula is by far the best-known "stellar nursery." It is visible with the naked eye as a hazy smudge of light below the three bright stars in Orion's Belt; yet it was first noticed in the seventeenth century after the invention of the telescope. Christiaan Huygens, in 1659, was first to publish a sketch of the nebula. In 1771, French astronomer Charles Messier included it as number 42 on his list of nebular objects (M42).

The Orion Nebula is part of the gigantic Orion Complex. It has a diameter of about 25 light-years and is 1,350 light-years from Earth. In the center of the nebula, four young, hot stars pump high-energy ultraviolet radiation into space. These four "Trapezium stars" keep their immediate surroundings clean by blasting away all material and generate shock waves and density enhancements in the gas and dust clouds around them. New stars are then formed from the enhancements. In this way, the process of star formation spreads in all directions like a blazing forest fire.

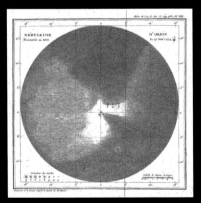

▲ Eighteenth-century astronomers such as French comet hunter Charles Messier made detailed pencil drawings of nebulae.

▲ This false-color infrared view of the Orion Nebula is based on observations of NASA's Spitzer Space Telescope and the European Herschel Space Observatory.

▲ In the heart of the Orion Nebula is the Trapezium, a cluster of four young, bright stars that energize their surroundings.

▶ The bright inner part of the Orion Nebula is known as the Huygens Region, after Dutch astronomer Christiaan Huygens, who was the first to draw and describe it.

▼ A huge mosaic of Hubble Space Telescope photographs showcases the splendor of the Orion Nebula, a stellar nursery at a distance of 1,350 light-years.

A Witches' Cauldron of Gas and Dust

PASSPORT

Name: Carina Nebula NGC 3372

Constellation: Carina

Sky position:
R.A. 10ʰ 45ᵐ 09ˢ
Dec. -59° 52.1'

Star chart: 11

Distance:
7,000 light-years

Diameter:
600 light-years

In the mid-eighteenth century, French astronomer Nicolas Louis de Lacaille travelled to the Cape of Good Hope to study the night sky in the southern hemisphere in detail. There, in 1751, he discovered a large, distinct nebular spot in the constellation Carina. The Carina Nebula is a gigantic star-forming region—much larger than the Orion Nebula—at least 7,000 light-years away in the Carina-Sagittarius spiral arm of the Milky Way. The nebula already contains a large number of young stars, including Eta Carinae, one of the heaviest stars in the Milky Way. In 1841, Eta Carinae experienced a catastrophic eruption and was, for a short time, the second brightest star in the sky, despite its enormous distance.

Large telescopes in the southern hemisphere have made detailed images of the Carina Nebula, and the Hubble Space Telescope has also studied the cosmic nursery very closely. You cannot take your eyes off the Hubble panorama, a tumultuous three-dimensional witches' cauldron of gas wisps, shock waves, dust clouds, and newborn stars. Some parts of the nebula seem to come straight out of fairy tales; one glowing protuberance has been christened "Mystic Mountain."

Thousands of stars have been born in the Carina Nebula over the course of millions of years; at various places in the nebula are young, compact star clusters. Other protostars are still embedded in dark clouds of dust and can be seen clearly only on infrared photos. These images also show the jets of hot gas that the protostars expel into space in two diametrically opposed directions. Who knows, perhaps another hypergiant like Eta Carinae will be born someday.

▼ Vaguely resembling an abstract James Pollock painting, this Hubble mosaic of the Carina Nebula is studded with loops of glowing gas and small, dark nebulae from which new stars will someday emerge.

▼ The European Very Large Telescope in Chile captured this detailed infrared image of the Carina Nebula.

◄ Located in the southern constellation of Carina, the bright Carina Nebula is invisible from mid-northern latitudes. This photograph was taken at the European Southern Observatory in Chile.

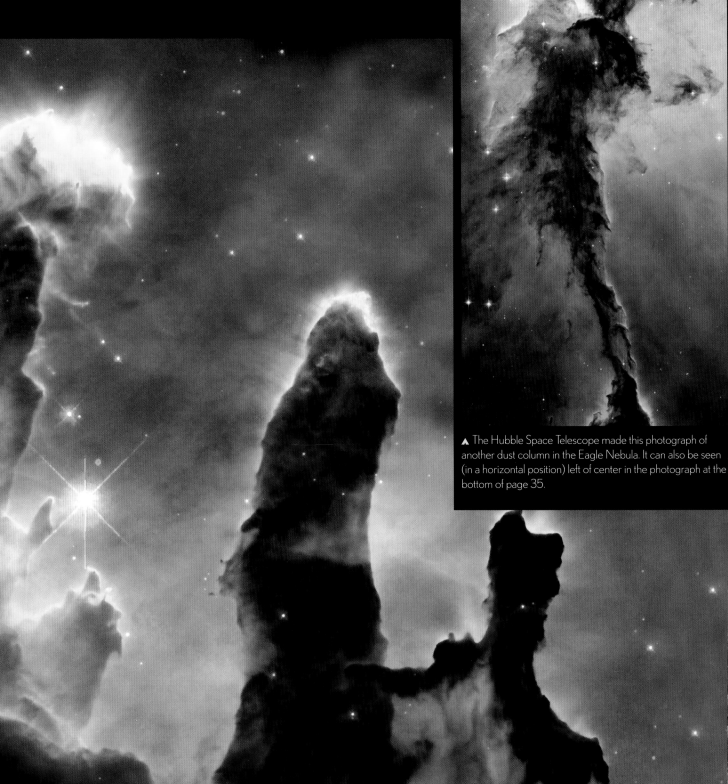

> Hubble's "Pillars of Creation" photograph of dark, eroding columns of dust in the Eagle Nebula made headlines in November 1995.

▲ The Hubble Space Telescope made this photograph of another dust column in the Eagle Nebula. It can also be seen (in a horizontal position) left of center in the photograph at the bottom of page 35.

Pillars of Dust in an Eagle's Nest

The Eagle Nebula in the constellation Serpens was front-page news in November 1995. The recently repaired Hubble Space Telescope had made detailed images of the central part of the nebula. The mysterious yellow-green photograph showed impressive "pillars" of dust that stood out sharply against the glowing background of the nebula. Small, dark protuberances were visible on the edges of the pillars—the birthplaces of new stars. The largest pillar was soon nicknamed the Finger of God, and the dark clumps were dubbed "eagle's eggs."

The Eagle Nebula (M16), around 7,000 light-years from Earth, had of course been discovered much earlier. Like the Orion Nebula, it glows because of the high-energy ultraviolet radiation of newborn stars at its center. This star cluster is only one to two million years old and contains some 450 stars; but the effect of the radiation on the surrounding gas and dust in the nebula had never been imaged in such detail.

The pillars of dust were formed in a similar way to the capricious sandstone formations in Utah, in the United States, or in Cappadocia, Turkey, except that here the erosion is caused not by wind or water but by radiation. At the tip of the Finger of God, that process of "photo evaporation" can be seen as a blue-green aura. The dark protuberances, which all point in the direction of the central star cluster, have such a high density that they do not allow themselves to be eroded away so easily; in some tens of thousands of years, new stars will be visible here.

PASSPORT

Name: Eagle Nebula
M16, NGC 6611

Constellation:
Serpens Cauda

Sky position:
R.A. 18ʰ 18ᵐ 48ˢ
Dec. -13° 49.0'

Star chart: 13

Distance:
7,000 light-years

Diameter: 20 light-years

▲ At infrared wavelengths, the famous pillars become almost transparent, revealing embedded protostars as well as background objects.

► The Eagle Nebula in the constellation Serpens Cauda looks like a flying bird of prey when seen through a small telescope.

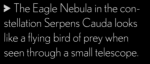

035

The Birth of Stars

PASSPORT

Name: Rosette Nebula
Caldwell 49

Constellation:
Monoceros

Sky position:
R.A. 06ʰ 33ᵐ 45ˢ
Dec. +04° 59.9'

Star chart: 4

Distance:
5,000 light-years

Diameter:
130 light-years

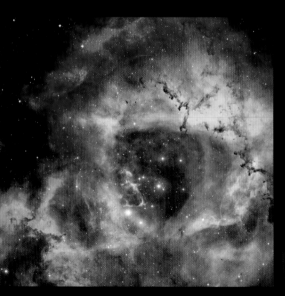

▼ Like the Orion and the Eagle Nebulae, the Rosette Nebula is energized by the ultraviolet radiation of a cluster of young, hot stars in the very center.

An Infrared Rosette

The European space telescope Herschel is named after English astronomer William Herschel, who discovered infrared radiation ("heat radiation") in the year 1800. The telescope was launched in the spring of 2009. For 4 years, its sensitive cameras and spectrographs recorded far-infrared and submillimeter radiation from the Universe. As this radiation is absorbed by water vapor in the atmosphere, it can best be observed from space.

Infrared telescopes are ideal for studying the birth of stars. Dark dust clouds, in which new stars are formed, are often difficult to see with normal telescopes, but they do emit heat radiation. Newborn stars that are still enclosed in their cocoon of gas and dust can be imaged using an infrared telescope, as the long-wavelength radiation passes right through the dust clouds.

To produce an image of the invisible infrared and submillimeter radiation, astronomers use "false colors." On this photograph of the Rosette Nebula in the constellation Monoceros—a large star-forming region at a distance of 5,000 light-years—radiation with a wavelength of 70 micrometers is shown in blue, 160 micrometers in green, and 250 micrometers (a quarter of a millimeter) in red.

To the right (just outside the image) is a young star cluster that has already formed at the center of the nebula. As in the Eagle Nebula, the high-energy radiation of the stars creates elongated pillars of dust in the surrounding nebula. The white specks in the center of the image are newly discovered baby stars. Some are ten times more massive than our own Sun. The Rosette Nebula contains sufficient material to form tens of thousands of suns.

➤ A small portion of the Rosette Nebula, as seen by the European infrared Herschel Space Observatory. At infrared wavelengths, dust-enshrouded protostars become visible.

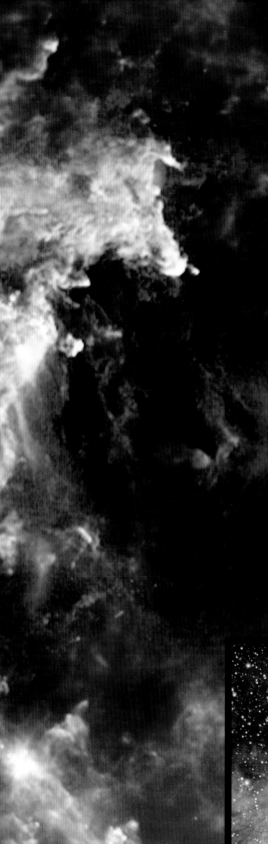

Star Factory

If the Tarantula Nebula were at the same distance from Earth as the Orion Nebula, it would never get dark at night. In reality, however, this colossal star-forming region is 160,000 light-years away, in the Large Magellanic Cloud—a small companion of the Milky Way. The nebula is visible to the naked eye, even at that great distance. Around 600 light-years in diameter, the Tarantula Nebula is one of the largest cosmic nurseries known to us.

The central part of this nebula, known as 30 Doradus, is a gigantic star factory. Star clusters several millions of years old can be found everywhere in the nebula. In the heart of 30 Doradus lies R136, a star cluster 35 light-years across and containing sufficient material to form almost half a million stars. Some of the newborn stars are hypergiants more than a hundred times more massive than the Sun.

The photo on page 26 is a mosaic of images made by the Hubble Space Telescope. It shows clouds of incandescent gas, dark wisps of dust, and countless twinkling newborn stars.

Interesting fact: the light from the Tarantula Nebula has taken 160,000 years to reach us. That means we see the nebula as it was when *Homo sapiens* first walked on Earth. Many of the massive stars we see now will in the meantime have exploded as supernovae, and many of the protostars will have grown to maturity.

PASSPORT

Name: Tarantula Nebula
NGC 2070

Constellation: Dorado

Sky position:
R.A. 05ʰ 38ᵐ 38ˢ
Dec. -69° 05.7'

Star chart: 14

Distance:
160,000 light-years

Diameter:
600 light-years

▼ In the center of this infrared image of the Tarantula Nebula is the isolated giant star VFTS 682, which weighs 150 solar masses or more.

The Birth of Stars

COSMIC NURSERIES

Star-forming regions come in a wide variety of shapes and sizes, but almost without exception they are of stunning beauty, especially when the glowing nebulae are sparkling with the blue-white light of newborn star clusters and when gaseous filaments are crisscrossed by dark strands of dust. These two pages present a colorful showcase of cosmic nurseries.

▲ A ridge of dark dust in a hot nebula in the Small Magellanic Cloud.

▼ The Lagoon Nebula is some 5,000 light-years away in the constellation Sagittarius.

▼ NGC 6559 is a small nebula, only a few light-years across.

▶ M17 in Sagittarius is known as the Omega or Swan Nebula.

▼ The Trifid Nebula is crisscrossed by dark ribbons of dust.

➤ RCW 108 is a giant star-forming region in the southern constellation Ara.

▲ N90 is a "spooky" stellar nursery in the Small Magellanic Cloud.

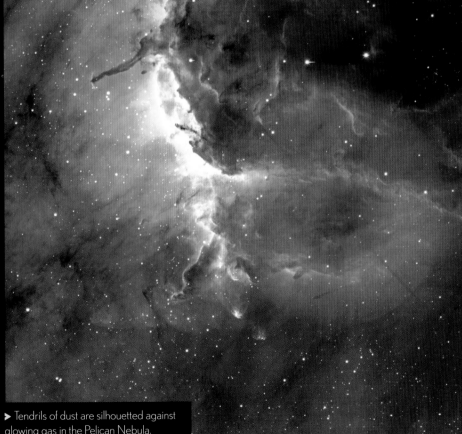

▲ The Cone Nebula (bottom) is at the apex of the Christmas Tree Cluster.

➤ Tendrils of dust are silhouetted against glowing gas in the Pelican Nebula.

A Family Affair

Stars are rarely born alone. Dark molecular clouds in the Milky Way usually contain sufficient gas and dust to form several thousand stars. In large star-forming regions, complete star clusters form, often containing hundreds of individual stars. These are known as open star clusters: the separate stars can be seen clearly with a telescope and one can, in effect, look right through the cluster.

The high-energy radiation of newborn stars in an open star cluster causes the gas in the surrounding nebula to glow. That happens, for example, in well-known cosmic nurseries like the Orion Nebula, the Eagle Nebula, and the Rosette Nebula. By no means are all young open star clusters that easy to see, however; in many cases, they are largely concealed from sight by the interstellar dust clouds in which they form and can be imaged only by an infrared telescope.

More than a thousand open star clusters have now been found in our own Milky Way. Some are only a few million years old, whereas others date back hundreds of millions of years. In astronomical terms, however, they are all young objects. As a result of gravitational disturbances, the individual stars gain sufficient speed to escape the cluster. Like children who grow up and leave their parental homes, the stars disseminate throughout the Universe, and the cluster gradually disintegrates.

Open star clusters are interesting research objects. The individual stars are all the same age, making it easier to compare their different characteristics. That has provided astronomers with a great amount of information about the life-span of stars.

▲ Using the European Southern Observatory's Visible and Infrared Survey Telescope for Astronomy (VISTA), astronomers discovered dozens of star clusters hidden behind galactic dust.

▶ Thousands of massive stars lurk in the core of the young star cluster NGC 3603 at a distance of about 20,000 light-years.

Young and Wild

PASSPORT

Name: NGC 3603

Constellation: Carina

Sky position:
R.A. 11ʰ 15ᵐ 09ˢ
Dec. -61° 16.3'

Star chart: 14

Distance:
20,000 light-years

Diameter: 5 light-years

Age: 2 million years

Around one million years ago, at a distance of about 20,000 light-years from Earth, an interstellar cloud of gas and dust collapsed under its own weight. In a short time, the cloud fragmented into many hundreds of individual "clumps," each of which eventually ignited as a newborn star. The result of this local "baby boom" was the young open star cluster NGC 3603, which was already discovered in 1834 by John Herschel, during his visit to South Africa. Misled by the fact that NGC 3603 is so compact, he thought it was a globular star cluster.

NGC 3603 has been studied extensively, including with the Hubble Space Telescope. It is in the Carina spiral arm of the Milky Way and is only a few light-years across. The most massive stars in the cluster are concentrated close to the center. Some are more than a hundred times more massive than the Sun. Such cosmic heavyweights have a very short life span; one of them is even about to explode as a supernova.

Hubble photographed the core of the star cluster in 1997 and again in 2007. By comparing the two images closely, the velocity of hundreds of stars could be measured. The measurements show that the star cluster has not yet really come to "rest"; the velocities of the individual stars are not related to their mass but nonetheless reflect the movements in the gas and dust cloud from which they formed a million years ago.

▼ Using Hubble Space Telescope images of the central part of the star cluster in the background image, astronomers have been able to measure the motions of hundreds of stars.

Seven Old Sisters

The Pleiades are the best-known open star cluster in the night sky. They have attracted the attention of everyone who looks at the stars since time immemorial. They are even mentioned in the Bible. This small but distinct "cloud" of stars in the constellation of Taurus was known to all ancient cultures and was recorded as early as prehistoric times in the caves of Lascaux and on the Nebra Sky Disk.

Despite its nickname, the Seven Sisters, the star cluster contains only six bright stars (Alcyone, Atlas, Electra, Maia, Merope, and Taygeta), but anyone with good eyesight can discern eight or ten, and many dozens can be seen with binoculars. If you also include the faintest members, you can count more than 1,000 stars, all the same age—more than 100 million years. The Pleiades are a relatively old star cluster.

The cloud of gas and dust from which the Pleiades were formed dissolved into space a long time ago. The wisps of nebulosity that can be seen around the brightest stars are part of another dust cloud through which the star cluster happens to be passing. The cold dust particles reflect the predominantly blue light of the brightest stars.

The Pleiades are only 430 light-years from Earth. Even closer, at a distance of 151 light-years, is a much older open star cluster, the Hyades. The cluster lies to the southwest of Aldebaran, the main star in the constellation Taurus. With an age of more than 600 million years, the Hyades are hardly recognizable as a star cluster anymore.

▲ The German Nebra sky disk, dating from the sixteenth century BC, probably features the Pleiades to the upper left of the lunar crescent.

▲ Interstellar dust in the Pleiades, heated by starlight, emits infrared radiation.

► Is the pattern of dots close to the bull's head in this prehistoric cave painting in Lascaux, France, a representation of the Pleiades star cluster in the constellation Taurus the Bull?

► Only the six or seven brightest stars of the Pleiades, or Seven Sisters, are visible to the naked eye. A large telescope is needed to note the faint nebulosity surrounding the Seven Sisters.

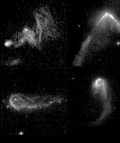

In a large star-forming region like the Orion Complex or the Carina Nebula, stars of all kinds and sizes are being born. Relatively small stars like our own Sun have a total life-span of many billions of years, but massive giant stars live for a much shorter time. Although they have greater reserves of nuclear fuel, they consume it at such a high speed that, within a few million years, they explode as supernovae, hardly giving these giant stars time to leave the area in which they were born.

Yet, in the mid-twentieth century, a few massive stars were discovered that move through the Milky Way in complete isolation and at extremely high speeds of more than 100 kilometers per second. In 1954, Dutch astronomer Adriaan Blaauw discovered that three of these stars—Mu Columbae, AE Aurigae, and 53 Arietis—originally come from the Orion Complex. They were apparently born there but escaped from their "parental home" while still very young and at extreme speeds.

A few years later, Blaauw presented a convincing explanation for this runaway behavior. Massive giant stars are often part of a compact binary star system, in which two stars orbit each other at high speed. If one of the twins explodes as a supernova, losing a large part of its mass, the other flies off into space like a stone from a sling.

Some runaway stars are accompanied by the small, compact neutron star left over after their companion exploded. When this was discovered at the end of the 1990s, it provided resounding proof of Blaauw's theory.

▲ Adriaan Blaauw's sketch of the motions of Mu Columbae, AE Aurigae, and 53 Arietis, projected backward, to show that they have a common origin in the Orion Complex.

▼ Energetic radiation from the star AE Aurigae heats up surrounding gas and dust in the Flaming Star Nebula, as seen in this infrared image by NASA's Wide-field Infrared Survey Explorer (WISE) space telescope.

➤ The bow shock surrounding the giant X-ray-emitting binary star Vela X-1, which contains a supernova remnant, reveals that it races through space at around 90 kilometers per second.

Pitch-Black Clouds

In *The Black Cloud*, a science fiction book by British astronomer Fred Hoyle, published in 1957, the solar system is visited by a dark cosmic cloud that possesses intelligence and is able to communicate. Hoyle described the compact cloud as a Bok globule, in honor of the Dutch-American astronomer Bart Bok, who had studied these small, pitch-black clouds in the Milky Way 10 years earlier.

Bok was convinced that his "globules" represented the final phase in the birth of a star. A small cloud of cool gas and dark dust contracts under the influence of its own gravity. Of course, one can see these small, dark clouds only when they stand out against a brighter background of glowing hydrogen gas or if they absorb the light of more distant stars. Inside the globule, a new star is formed; once it has produced enough energy, it will blow all the material away from its immediate surroundings and the globule will dissolve.

In the middle of the previous century, many astronomers did not take Bok's theory seriously. But in the 1980s, the Dutch-American Infrared Astronomical Satellite (IRAS) discovered that there is indeed a weak source of heat radiation at the center of nearly all Bok globules: an embryonic star.

On the photo of the Bok globule Barnard 68, it can clearly be seen that the density of the darker cloud increases toward the center. The background stars on the edge of the cloud are still visible, but their light is strongly reddened by the absorbing effect of the dust, just as the light of the setting Sun turns red from being absorbed by the atmosphere.

PASSPORT

Name: Barnard 68

Constellation: Ophiuchus

Sky position:
R.A. 17ʰ 22ᵐ 38ˢ
Dec. -23° 49.6'

Star chart: 12

Distance: 500 light-years

Diameter: 0.5 light-years

▲ Looking like blotches of spilled ink, dark globules are seen silhouetted against the glow of hot hydrogen gas in IC 2944, a stellar nursery in the constellation Centaurus.

➤ Barnard 68 is a well-known Bok globule in the constellation Ophiuchus.

PASSPORT

Name:
Horsehead Nebula
Barnard 33

Constellation: Orion

Sky position:
R.A. 05ʰ 40ᵐ 59ˢ
Dec. -02° 27.5'

Star chart: 9

Distance:
1,500 light-years

Diameter: 3 light-years

Cosmic Whinnying

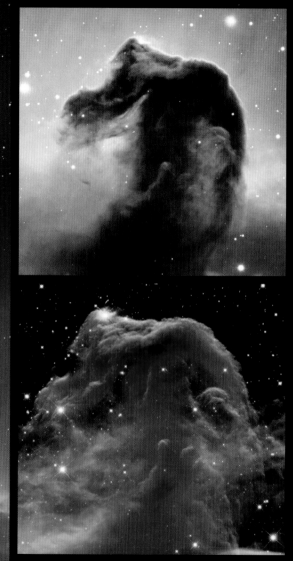

One needs an active imagination to recognize a great bear, a bull, or a dragon in the constellations; but some nebulae really do justice to their names. In 1888, Scottish astronomer Williamina Fleming was studying a photographic plate of the constellation Orion made at the observatory at Harvard University. Just below the bright star Alnitak, she saw a remarkable small, dark nebula that closely resembled the head of a horse. The official name of the object is Barnard 33, but it has been known ever since as the Horsehead Nebula.

In Fleming's time, no one knew how stars produce light and heat, let alone how they are born from cosmic clouds of gas and dust. Today we know that the Horsehead Nebula is a large, irregularly shaped Bok globule, which appears dark against a pink background of glowing hydrogen gas. It is part of the expansive Orion Complex. In tens of thousands of years, new stars will shine here in the night sky.

The large photo of the Horsehead Nebula shows parallel pink wisps of gas, probably caused by magnetic fields. The dark nebula is a small protuberance of the large, dark cloud of cool hydrogen gas that fills the lower half of the photo. On the infrared image at the right, this extended cloud is clearly visible, but its horsehead form is less clearly defined because infrared radiation is not much affected by dust. The Hubble Space Telescope captured a detailed infrared image of the nebula; several embryonic stars can be seen on the photo.

▲ The Horsehead Nebula is at the lower right in this infrared image of the Flame Nebula, made by the European Visible and Infrared Survey Telescope for Astronomy (VISTA). The star Alnitak is at the top right.

◄ A delicate interplay of bright glowing gas and dark-absorbing dust creates the illusion of a horse's head protruding from a cloud deck.

▲ At visible wavelengths (top), the Horsehead Nebula is an impenetrable dark cloud. At near-infrared wavelengths (bottom, Hubble image), it becomes much more transparent.

047

The Bir

"We Are Pleased to Announce..."

Stars are not born from one day to the next. A star like the Sun can live to be 10 billion years old—more than a 100 million times the lifespan of a human being. So it is not surprising that the birth of a star also takes much longer, easily at least 100,000 years.

Molecular clouds and glowing nebulae of hydrogen gas are the nurseries in which new stars are born. The small, dark Bok globules could be described as cosmic wombs. When the stellar fetus is ready to come into the world, it is called a *protostar*.

Protostars are difficult to see. They are usually still hidden away in the dark clouds of gas and dust in which they are formed, but infrared telescopes can detect their faint heat radiation. In this way, astronomers have discovered that they are contracting spherical masses of gas. They radiate energy because they are contracting under their own gravity.

Apart from infrared radiation, radio waves, and X-rays, newborn stars also emit electrically charged particles into space. This stellar wind, which is particularly strong along the rotational axis of the protostar, produces shock waves and density enhancements in the surrounding nebular gas. These small patches of nebulosity are known as Herbig-Haro objects after the two astronomers who discovered them. Herbig-Haro objects are the first "cries" of a newborn star.

Eventually, the temperature and density in the interior of the protostar are high enough to cause spontaneous nuclear reactions. The energy released during these reactions brings the contraction of the protostar to a halt. Only when that new equilibrium has been reached can the newborn be called a real star.

▲ Using the infrared vision of the European Southern Observatory's Very Large Telescope, astronomers discovered a handful of protostars in the core of the star-forming region RCW 38.

▲ A young protostar spews jets of matter into space in this image of HH 47, obtained with the Hubble Space Telescope.

Embryonic Planets

Nothing is perfect in nature. When a cosmic cloud of gas and dust condenses to form a new star, not all the material ends up in the star. A small part of the cloud—usually no more than a few percent—continues to orbit the star in a flattened, rotating disk known as a *protoplanetary disk*, or *proplyd* for short. With a little luck, the material in the disk will clump into one or more planets.

That our own planetary system was formed from a flat, rotating disk was suggested as early as the mid-eighteenth century by the philosopher Immanuel Kant; but it was not until the middle of the 1980s that the American-Dutch infrared satellite IRAS found evidence of the existence of circumstellar disks, for example, around the stars Beta Pictoris and Vega. In the early 1990s, the Hubble Space Telescope discovered protoplanetary disks in the Orion Nebula. Since then, they have been found in many other star-forming regions.

How can a contracting cloud of gas and dust produce a flat, rotating disk of material? The answer is simple: even the slightest coincidental rotation of the cloud is accelerated by the process of contraction, in the same way that an ice-skater spins more and more quickly during a pirouette as she moves her arms in closer to her sides (the law of conservation of angular momentum). Everything that rotates automatically becomes flatter as a result of centrifugal force. An example is seen in a pizzeria, where a lump of dough thrown up in the air with a rotating motion comes back down as a flat pancake.

▲ An artistic impression of young stars and circumstellar protoplanetary disks in a stellar nursery.

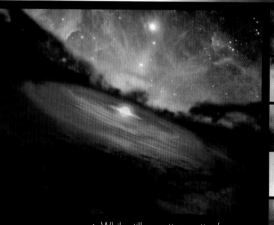

▲ While still accreting matter from a surrounding disk of gas and dust, a young protostar blows jets of material into space along its rotational axis.

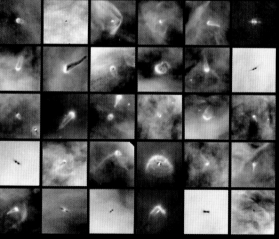

▲ Dozens of protoplanetary disks have been discovered in the Carina Nebula, a giant star-forming region at a distance of about 7,000 light-years.

▼ One of the first images of a protoplanetary disk, captured by the Hubble Space Telescope in the Orion Nebula.

Fomalhaut's Family

Fomalhaut is one of the brightest stars in the night sky. It radiates nearly seventeen times as much light as the Sun and is also quite close—only 25 light-years away. Infrared telescopes had already discovered that this star, in the constellation Piscis Austrinus (the Southern Fish), must be surrounded by an enormous flattened ring of gas and dust. The Hubble Space Telescope was the first to image this protoplanetary disk.

Fomalhaut is a hot, young star. It was born a few hundred million years ago, probably together with the bright stars Castor and Vega. It may also be accompanied by a few massive planets; the sharp edges of the dust ring could be the result of gravitational disturbances caused by these planets.

In 2008, American astronomers announced the discovery of a moving object in the Fomalhaut dust ring. In the years since then, the movement of the bright "speck" has been clear to see. Perhaps it is a massive, Jupiter-like planet. It has been given the provisional name Fomalhaut b.

Not everyone is convinced that Fomalhaut b is a real planet. Measurements with the Spitzer Space Telescope suggest that it may be a relatively compact dust cloud, perhaps with an embryonic planet at its center. There is no doubt at all, however, that many kinds of objects are orbiting Fomalhaut. The European space telescope Herschel has discovered particles with dimensions of around 1 micrometer that may have been formed as the result of collisions between comets.

> Ploughing through the disk of gas and dust surrounding Fomalhaut (upper left), a young giant planet is pictured in this artistic impression.

◄ Fomalhaut is the brightest star in the constellation Piscis Austrinus, the Southern Fish.

◄ Larger dust particles in Fomalhaut's disk are concentrated in a narrow ring, as seen in this image by the international Atacama Large Millimeter/Submillimeter Array (ALMA) observatory in Chile.

➤ A bright spot, called Fomalhaut b, can
be seen orbiting in the protoplanetary
disk surrounding Fomalhaut. Could it be
a planet?

The Birth of Stars

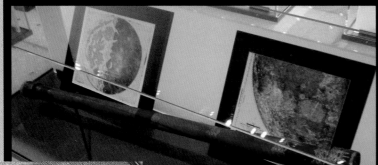

Telescopes

The telescope was invented around the year 1600 by Hans Lippershey and Zacharias Janssen, who were spectacle-makers in the Netherlands. The principle is simple: a convex lens (the objective) produces a detailed image of the observed object, which can then be looked at using a kind of magnifying glass (the eyepiece). In 1688, Isaac Newton devised the first reflecting telescope, in which the objective is a convex mirror.

With both refracting (lens) and reflecting telescopes, a larger objective will show fainter stars and more details. It is therefore not surprising that over time astronomers built increasingly large telescopes, like the 1.2-meter telescope designed by William Herschel in 1789. The 1.8-meter Leviathan, built for Irish astronomer William Parsons (Lord Rosse) in 1845, was the largest in the world for more than 70 years.

In the first half of the twentieth century, large telescopes became too expensive to be financed by private individuals. American astronomer George Ellery Hale found a number of wealthy businessmen willing to sponsor the construction of several successors to Lord Rosse's Leviathan, including the 2.5-meter Hooker Telescope on Mount Wilson (1918), California, which was used to discover the expansion of the Universe, and the 5-meter Hale Telescope on Palomar Mountain (1948), also in California.

After the Second World War, astronomers built the first radio telescopes, large parabolic dishes that can detect faint radio waves from the cosmos. Because radio waves have a much longer wavelength than visible light, the design requirements for radio telescopes, such as surface accuracy, are less stringent. They can therefore be much larger than optical telescopes. The largest radio dish in the world, with a diameter of 305 meters, is at Arecibo Observatory in Puerto Rico.

▲ Galileo Galilei used this homemade telescope in 1609 to make detailed drawings of the moon.

▲ In 1608, Dutch spectacle-maker Hans Lippershey was the first to officially describe the invention of the telescope.

▶ William Herschel's giant 40-foot telescope was the largest in the world at the end of the eighteenth century.

▶ Aerial view of the Very Large Array radio telescope in Socorro, New Mexico.

◀ The Great Refractor at Yerkes Observatory in Williams Bay, Wisconsin, is the largest refracting telescope in the world.

▲ At Birr Castle in Ireland in 1845, William Parsons oversees the construction of his 1.8-meter "Leviathan" telescope in this drawing by Herbert Crompton Herries.

▲ Thousands of radio dishes and smaller antennae will make up the future Square Kilometre Array, which will be constructed in Australia and southern Africa.

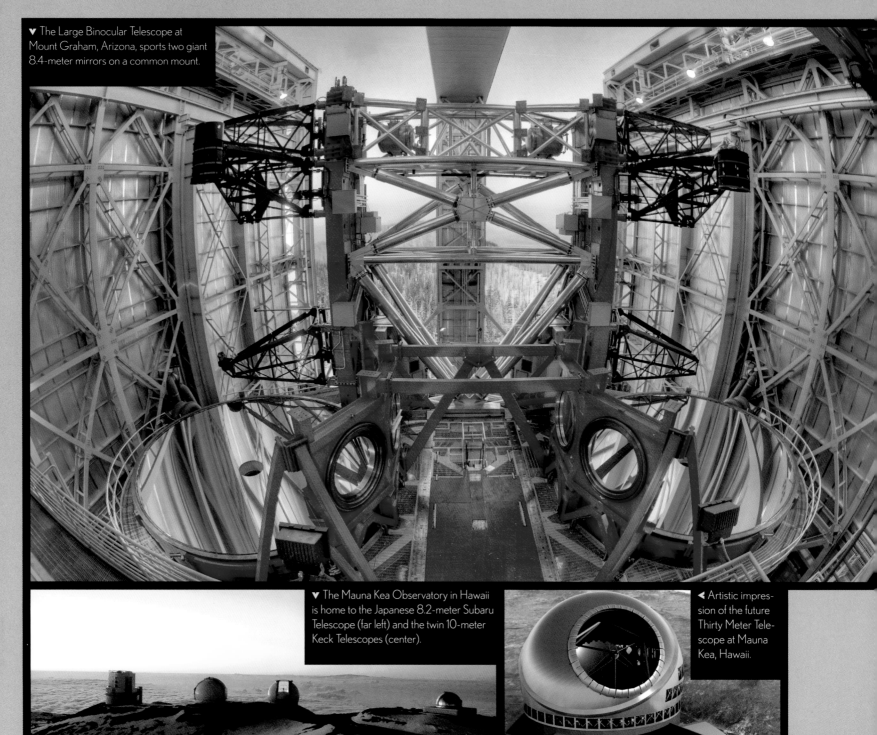

▾ The Large Binocular Telescope at Mount Graham, Arizona, sports two giant 8.4-meter mirrors on a common mount.

▾ The Mauna Kea Observatory in Hawaii is home to the Japanese 8.2-meter Subaru Telescope (far left) and the twin 10-meter Keck Telescopes (center).

◄ Artistic impression of the future Thirty Meter Telescope at Mauna Kea, Hawaii.

Deep Space

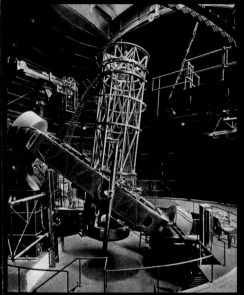

Since the 1980s, astronomers have been building increasingly large optical and infrared telescopes. "Active optics" enable the relatively thin mirrors (which can be up to 8.4 meters in diameter) to be continually adjusted to compensate for sagging and for the effects of wind load. Even larger mirrors, up to 10 meters across, are no longer made in one piece but consist of dozens of hexagonal segments, a method first applied with the two Keck telescopes of the W. M. Keck Observatory on Mauna Kea, Hawaii.

Every present-day giant telescope is equipped with "adaptive optics." The disturbing effects of vibrations in the atmosphere are neutralized by measuring them 100 times a second—often using laser beams—and passing the signals on to a "flexible" auxiliary mirror in the light path, which is then made to vibrate in such a way that the resulting image is again completely still.

Interferometry, which is used, for example, in the European Very Large Telescope in Chile (European Southern Observatory), and the Large Binocular Telescope in Arizona, makes it possible to combine observations by different telescope mirrors to achieve an extremely sharp image. The same technique has already been used in radio astronomy for more than a century.

In the near future, there are plans to build monster telescopes with composite or segmented mirrors of 20 to 40 meters, including the Giant Magellan Telescope (Chile), the Thirty Meter Telescope (Hawaii), and the European Extremely Large Telescope (Chile). Radio astronomers are also currently working on the design of the Square Kilometer Array, tens of thousands of linked dishes and antennae in Australia and South Africa with a total antenna surface area of a square kilometer.

◄ The 100-inch Hooker Telescope at Mount Wilson, California, was used to discover the expansion of the Universe.

▲ The Grand Old Lady of twentieth-century telescopes is the 5-meter Hale reflector at Palomar Mountain, California, inaugurated in 1948.

▲ At the European Very Large Telescope at Paranal Observatory, Chile, sodium lasers are used to measure atmospheric turbulence, which can then be compensated for by using adaptive optics.

➤ At Cerro Armazones in northern Chile, work is under way for the construction of the European Extremely Large Telescope, with a segmented mirror 39.2 meters in diameter.

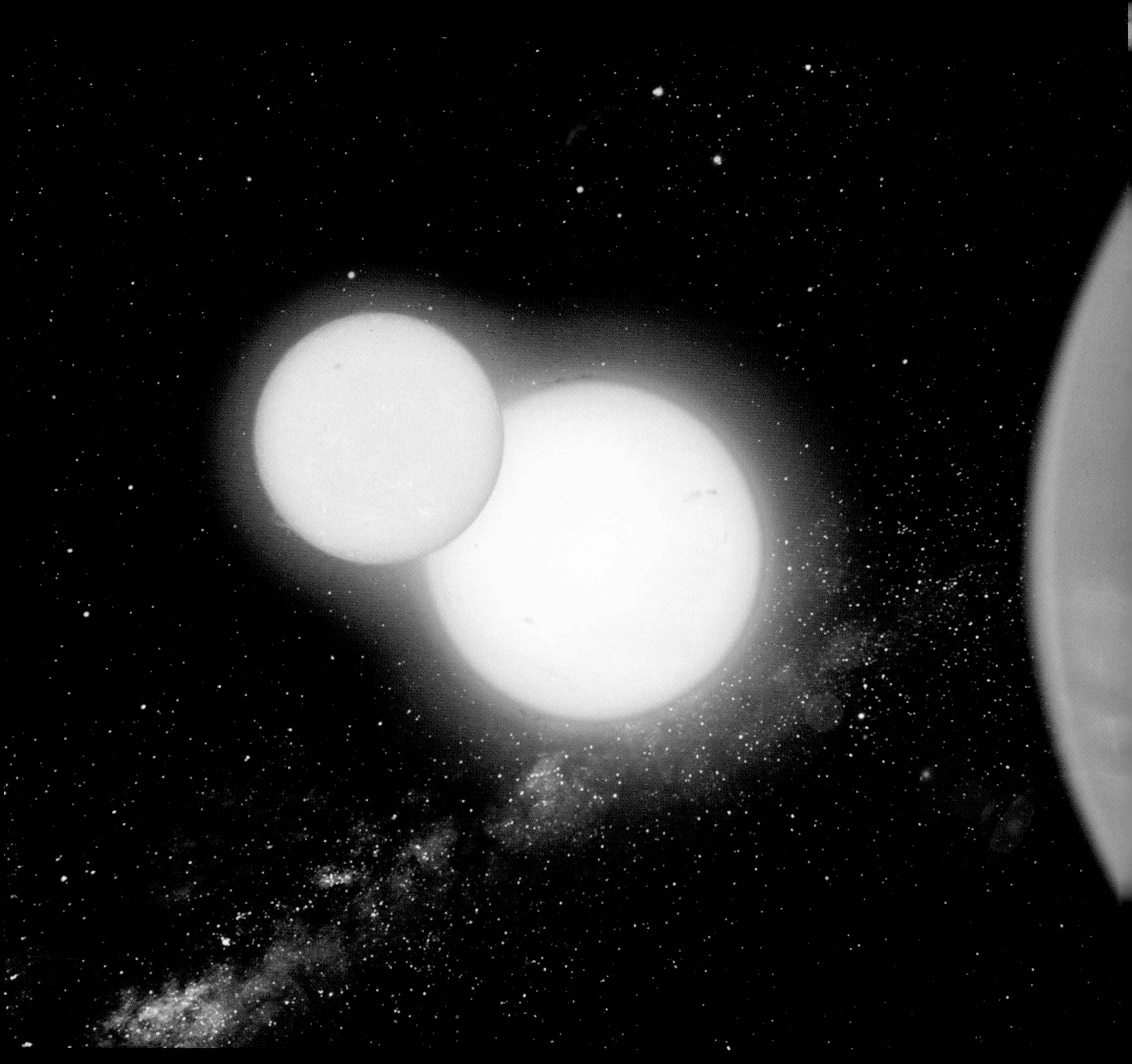

STARS and PLANETS

Each star one sees in the night sky is a gigantic ball of extremely hot incandescent gas, just like our own Sun. They are cosmic nuclear power plants, where energy generated as one element is transformed into another. In these nuclear witches' cauldrons, increasingly heavy atoms are created, which later form the building blocks of organic molecules and living organisms.

In 1835 the French philosopher Auguste Comte claimed that we would never be able to discover what stars are made of; after all, one cannot take a sample from a star. Not long afterward, however, the spectroscope was invented, enabling astronomers to study not only the composition and structure of stars but also how they are born, live, and die.

We now know that the cosmos is one large recycling plant in which interstellar clouds of gas and dust collapse under their own gravity to form stars with the most diverse properties. At the end of their lives, those stars blow a large part of their matter back out into space. From that stellar gas—enriched with newly formed heavy elements—new stars and planets can be formed.

The most spectacular development in astronomy in recent years is the discovery of exoplanets, planets orbiting other stars than the Sun. At least half of all stars are accompanied by one or more planets. These exoplanets also display an unimaginable diversity, but it is as good as certain that some of them will be almost identical twin sisters of our own Earth.

◄ Kepler-35b is a giant extrasolar planet that orbits a binary star at a distance of 5,400 light-years.

Cosmic Nuclear Reactors

A star is nothing more than a gigantic ball of incandescent hot gas. Roughly speaking, stars consist of about 75% hydrogen, the simplest and lightest element in the Universe. Most of the remaining quarter is helium, the second-lightest element. Most stars contain only small quantities of other, heavier elements. A star pumps energy out into space for hundreds of millions or even billions of years, but where does this energy come from?

The puzzle of where stars get their energy was not solved until the first half of the twentieth century, when physicists gained a greater insight into the structure of atoms and the nuclear reactions they can undergo. Because of the enormously high pressure and temperature in the interior of a star, hydrogen atoms can be so densely compressed that spontaneous nuclear fusion can occur. Small, light atomic nuclei fuse together to form larger, heavier nuclei. It is in effect a form of cosmic alchemy, with one element being transformed into another.

The most common form of fusion is that of hydrogen into helium. The details are quite complicated, but the net effect is that, after passing through several interim stages, four hydrogen atoms fuse to form one helium atom. This hydrogen fusion releases an enormous amount of energy, as anyone knows who has ever seen pictures of the devastating effect of a hydrogen bomb. Our Sun is, in a certain sense, a hydrogen bomb that has been exploding for the past 4.6 billion years. And each and every star in the night sky is a cosmic nuclear reactor.

▲ The hydrogen bomb was human-kind's unfortunate way to copy the Universe's process of energy production through nuclear fusion.

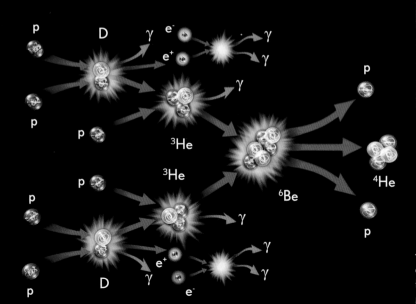

◄ In stellar nuclear fusion, through a number of intermediate steps involving deuterium (D) and beryllium (Be), four hydrogen nuclei (protons, p) fuse into one helium nucleus (He).

Ordering the Stars

Stars come in all shapes and sizes: large and small, hot and cool, bright and faint. These properties also occur in different combinations. Hot stars are not always bright, faint stars are not always small, and large stars can also be cool; but not every combination of properties occurs with equal frequency. The Hertzsprung-Russell diagram, named after the two astronomers that created it, shows the enormous variety of stars at one glance.

The horizontal axis shows the temperature. The surface temperature of a star determines its color, just as the color of an iron poker in a fire is determined by its temperature. Hot, blue-white stars are on the left of the diagram and cool, red stars on the right. The vertical axis shows the luminosity of the star, which is a way of measuring the energy it radiates. Low-luminosity stars are at the bottom of the diagram, and the most luminous ones are at the top. It is immediately clear that small stars should be at the bottom left and large stars top right; if a star is very hot yet faint, it must be small. Stars that are cool but very bright must be enormously big.

If all the stars are placed in their rightful position in the Hertzsprung-Russell diagram, you can see that most of them (including our own Sun) lie on a diagonal band running from the top left to the bottom right, from the blue giants to the red dwarfs. This band is known as the main sequence. Every star spends most of its life somewhere on the main sequence, namely, when hydrogen fusion is occurring in its interior. Red giants (top right) and white dwarfs (bottom left) are less common.

▼ In the Hertzsprung-Russell diagram, a plot of luminosity versus surface temperature (or spectral class), stars fall into a small number of discrete groups.

LUMINOSITY
(Solar Units)

1000000
100000
10000
1000
100
10
1
0.1
0.01
0.001
0.0001

Blue Supergiants

Red Supergiants

Massive Stars

Red Giants

Main Sequence

Sunlike Stars

White Dwarfs

Red Dwarfs

O   B   A F G K M

30,000
10,000
6,000
3,000

SPECTRAL CLASS &
SURFACE TEMPERATURE (Kelvin)

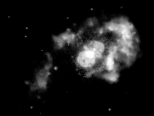

▲ NASA's Chandra X-ray Observatory captured the X-rays emitted by the hot gas surrounding Eta Carinae.

◀ The central 'Homunculus Nebula' of Eta Carinae was produced in a giant explosion in 1843, which also sent a blast wave through the surrounding gas.

▼ In the very center of the Tarantula Nebula is the giant star cluster R136, which contains some of the most massive stars known.

Stellar Heavyweights

The larger and more massive a contracting cloud of gas and dust, the larger and more massive the resulting star will be; that seems logical. Yet there is a limit to the size and mass of stars. When a star has more than 120 times the mass of the Sun (the Eddington limit), it produces so much radiation that it "blows itself up." The star's radiation pressure proves more powerful than its gravity.

Strangely enough, a number of exceptions to this rule have been discovered. The star Eta Carinae, for example, has 150 times more mass than the Sun and was probably 20% heavier when it was born. In the nineteenth century, Eta Carinae experienced a short series of spectacular explosions. In the spring of 1843, it was even nearly as bright as

Sirius, the brightest star in the winter sky, despite being nearly 1,000 times farther away.

R136a1, a star in the Large Magellanic Cloud, is even more extreme. It is 265 times more massive than the Sun and is almost nine million times more luminous. The surface temperature of this cosmic heavyweight is more than 50,000°. Like Eta Carinae, it blows enormous quantities of gas into space; in the past million years, it has lost the equivalent mass of perhaps fifty Suns.

How are these monster stars able to defy the Eddington limit? No one knows for sure, but they are found in densely populated, compact star clusters, so they may have been created by the collision and merger of two or more lighter stars.

Straggling Cannibals

High-mass stars have shorter lives than low-mass ones. A heavy star may have more mass, and therefore a greater supply of hydrogen, but it burns it up much more quickly. As a result, hydrogen fusion ceases after only a few tens of millions of years, and the star begins to evolve further. In the Hertzsprung-Russell diagram, it moves away from the main sequence. Lighter stars can sometimes go on for billions of years with their smaller supplies of hydrogen and therefore stay on the main sequence for much longer.

Astronomers were therefore astounded at the discovery of hot, massive stars that stayed on the main sequence much longer than expected, as if they were evolving at a much more leisurely pace. These blue giants are known as *blue stragglers*. They are most commonly found in compact star clusters.

The stars in a cluster are usually all born around the same time; and, in the Hertzsprung-Russell diagram of the cluster, it is evident that the most massive stars have already left the main sequence, except, that is, for the blue stragglers.

The explanation is as spectacular as it is simple. Star clusters contain many binary stars: two stars that circle around each other in a small orbit. When one of the twins swells up at the end of its life, it can swallow up its companion. It is as though the star is refueling, enabling hydrogen fusion to continue for much longer. A collision between two stars can also lead them to merge together.

Blue stragglers live longer by swallowing up their fellow stars. They are the cannibals of the cosmos.

▼ In dense star clusters like NGC 6397, interactions between individual stars can lead to the formation of "blue stragglers."

➤ Mass transfer or merger: there are two routes to the formation of a hot, blue star that looks younger than it really is.

PASSPORT

Name: 18 Scorpii

Constellation: Scorpius

Sky position:
R.A. 16ʰ 15ᵐ 37ˢ
Dec. -08° 22.2'

Star chart: 12

Distance: 45 light-years

Brightness: 5.5ᵐ

Mass: 1.02 x Sun

Diameter: 1.01 x Sun

Luminosity: 1.06 x Sun

Age: 4.6 billion years

Stars like the Sun

Our Sun is an average star, not especially large but not extremely small either; not too hot, but also not very cool; not particularly bright, but by no means the runt of the cosmic litter. It is the Milky Way's equivalent of "the guy in the street."

Because the Sun is such an average star, it means that there must be many stars like it in the Milky Way, around 1.5 million kilometers across, with a surface temperature of 5,000 or 6,000°C, an energy output of some 400 quintillion megawatts, and about five billion years old.

The star that most closely resembles the Sun is HIP 56948, a small, faint star in the constellation Draco, which is about 200 light-years away. If HIP 56948 were to take the place of the Sun tomorrow,

we would notice little difference here on Earth. 18 Scorpii, a star at a distance of only 45 light-years, and therefore just visible with the naked eye, is also a carbon copy of the Sun. We do not know yet whether it has any planets.

The study of Sun-like stars is very important in understanding our own star better. Only by comparing the Sun closely with its twin brothers and sisters is it possible to determine whether certain properties (such as the eleven-year activity cycle and the occurrence of long-term minima within it) are specific or generic. It has, for example, already been discovered that the energy production of the Sun is extremely stable compared with that of most other stars.

▼ Every 11 years, the number of dark sunspots and bright active regions on the Sun's luminous surface is higher than average.

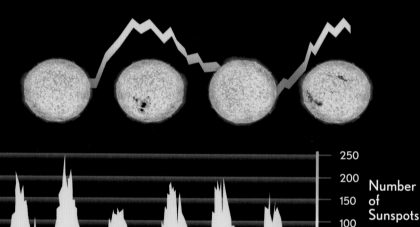

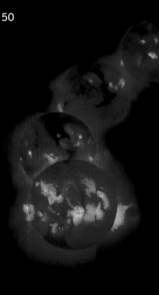

▲ The number of sunspots and sunspot groups is a measure of solar activity. Not every maximum has the same intensity.

➤ During a solar maximum (foreground), the Sun produces more high-energy X-rays than during a minimum (background).

062

Deep Space

▲ This Dutch winter scene, painted by Hendrick Avercamp, reflects conditions during the Little Ice Age, which coincided with an extended solar minimum between 1645 and 1715.

▲ The stars HIP 56948 (left) and 18 Scorpii (right) are carbon copies of our own Sun.

The Red Dwarf Army

In the natural world, there are always more small things than large ones. Look at the beach: there are a few large boulders, many more rocks and pebbles, and countless grains of sand. And so it is in the Universe: there are many more small, low-mass stars than large, massive ones.

Red dwarfs are by far the most common stars in the cosmos. There are a few hundred billion in the Milky Way alone. Red dwarfs are born small–they are not much larger than the giant planet Jupiter. The pressure and temperature in their interior are just about high enough to spark off hydrogen fusion reactions. The reactions do not occur quickly and, although red dwarfs have much smaller supplies of hydrogen than massive giant stars, they can make them last for many billions of years. Red dwarfs have almost eternal life.

It is not possible to see a single red dwarf with the naked eye, as these faint stars simply do not emit sufficient light. Yet, of the fifteen stars closest to the Sun, no fewer than ten are red dwarfs, showing just how numerous they are. The two most well known are Proxima Centauri (the nearest star, at a distance of 4.24 light-years) and Barnard's Star (about 6 light-years away).

Many red dwarfs are accompanied by one or more planets. It is, however, uncertain whether anything can live on these planets, as red dwarfs frequently experience eruptions of high-energy X-rays.

▲ To the right of Gliese 623 is the red dwarf star Gliese 623b, which is 60,000 times fainter than the Sun.

➤ The Hubble Space Telescope captured this image of the red dwarf Proxima Centauri, the nearest neighbor of the Sun.

➤ A red dwarf star is much smaller than the Sun and has a lower surface temperature of only a few thousand degrees Celsius.

▲ The detection of large dust particles in a disk surrounding a young brown dwarf shows that even these "failed stars" may grow rocky planets.

▲ The brown dwarf binary CFBDSIR 1458+10 has the same temperature as a cup of tea.

Failed Stars

A star is a real star only when energy is released in its interior through the fusion of hydrogen into helium. These nuclear fusion reactions occur when the pressure and temperature in the core of the star are high enough. For that to happen, the star must be at least eighty times as massive as Jupiter. But what happens if a minute interstellar cloud, with a mass lower than eighty Jupiters, collapses under its own weight?

American astronomer Jill Tarter calculated this event for her thesis in 1975. The very smallest clumps of hydrogen and helium gas (like Jupiter) do not have sufficient mass to give rise to spontaneous hydrogen fusion. However, Tarter discovered that if an object is more massive than thirteen Jupiters, deuterium fusion can take place. Deuterium (heavy hydrogen) occurs in small quantities in stars, and the nuclear fusion of deuterium also generates a small amount of energy.

Tarter called these "failed stars" *brown dwarfs*; after all, they do radiate a small quantity of energy (primarily in the form of infrared heat radiation), so they are not completely black. In reality, brown dwarfs probably have a magenta tint.

Although brown dwarfs are thirteen to eighty times more massive than Jupiter, they are not much larger than the giant planet. Their surface temperature is in some cases no higher than a few tens of degrees Celsius, and clouds may even form in their atmospheres. It is not surprising that very few brown dwarfs have been discovered so far, as they are almost impossible to observe.

Stars and Plane

◄ An artistic impression of a Y dwarf, the coolest type of brown dwarf. Y dwarfs are about the same size as the planet Jupiter.

▼ The most famous binary star in the sky can be found in the tail of the Great Bear: Mizar (the brightest component) and Alcor.

Spouses in Spac

On the fictional desert planet Tatooine, the home of Luke Skywalker from the famous *Star Wars* movies, there is a double sunset every evening. Tatooine does not orbit one star, but two. *Binary stars*—twin stars that go through their lives together—are common in the Universe: more than half of all stars are part of a binary system.

The best-known binary star in the sky is Mizar, the middle star in the tail of Ursa Major, the Great Bear. Anyone with good eyesight can see that Mizar is accompanied at a short distance by a fainter star, known as Alcor. Most other binary stars can be "separated" only by looking through a telescope. It is not possible, for example, to see with the naked eye that the closest star to Earth, Alpha Centauri, actually has a close companion. The bright star Castor, in the constellation Gemini, is part of a six-star system.

The reason there are so many binary and multiple star systems in the Universe is that, as it collapses under its own gravity, a protostellar cloud easily splits into fragments, which may lead to the birth of two stars with roughly the same mass, as is the case with Alpha Centauri. Sometimes, it can produce a binary star system with one large, massive star and a smaller, less massive twin, such as a red or brown dwarf.

When one of the two stars in a binary system swells up at the end of its life and becomes a red giant, material can be transferred to its companion. That can produce all kinds of spectacular phenomena, such as nova explosions.

▲ If they are close together, the two stars in a binary can affect each other's evolution.

➤ Luke Skywalker enjoys a double sunset on his home planet Tatooine, which orbits a binary star.

Cosmic Flashing Lights

Imagine that one day the Sun radiated one and a half times as much light and heat as on another day. That would have an enormous impact on our weather and climate. Life on Earth—if it were even possible under such circumstances—would be completely different. Fortunately for us, the Sun is extremely stable; but many other stars are much less constant, and their brightness changes over the course of hours, days, weeks, or months.

Some small stars, like red dwarfs, display irregular explosions during which large quantities of high-energy X-rays are released. Catastrophic explosions also occur from time to time in binary star systems, for example, when one star suddenly transfers a large quantity of material to the other. In the case of most variable stars, however, the changes in brightness are much more regular. They arise because the star pulsates, becoming alternately larger and smaller, so that its temperature and brightness also vary.

The best-known pulsating variable stars are known as Cepheids, after the prototype Delta Cephei. In the early twentieth century, American astronomer Henrietta Leavitt discovered that there is a strict relationship between the pulsation period of a Cepheid and its average luminosity: the brighter the Cepheid, the slower it pulsates.

Astronomers use Cepheids to determine the distances to other galaxies. Quite simply, they measure the pulsation period and, applying Leavitt's law, calculate the average luminosity of the star. By comparing that with its observed brightness in the night sky, one can easily calculate how far away it is.

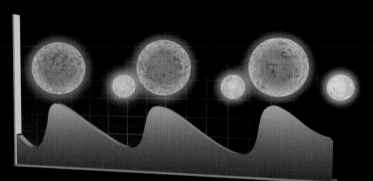

▲ The variable star Mira, at the far right of this ultraviolet photograph, leaves a trail of hot gas as it races through space.

◄ Cepheid variable stars brighten and dim as they pulsate. The pulsations are slower for more luminous Cepheids.

▼ Observations of Cepheids in the spiral arms of galaxy M100 have enabled astronomers to measure the galaxy's distance.

Stars and Planets

> The white dwarf star (on the right) in the RS Ophiuchi binary system has just exploded as a nova in this artistic impression.

▼ A small expanding nebula was blown into space in 1901, when Nova Persei erupted.

▲ Matter from a companion star piles up on the surface of a white dwarf—until a thermonuclear chain reaction takes place.

Going Nova

In 1975, amateur astronomers around the world saw a new star appear in the constellation Cygnus. Nova Cygni 1975 (the Latin word nova means "new") was one of the brightest newcomers to appear in the night sky in recent decades. In reality, however, it was not a new star at all: novae are powerful explosions of stars that are normally so faint that one can hardly see them, if at all.

Nova explosions occur in binary star systems that have already been evolving for a long time. Two more or less Sun-like stars orbit each other. The more massive of the two evolves the quickest. At the end of its life, it swells up to become a red giant, expels its outer gas layers into space, and ends up as a small, compact white dwarf—more massive than the Sun but hardly larger than Earth.

Later, the companion also swells up to become a red giant. Material from this star, mostly hydrogen and helium, is attracted by the white dwarf and accumulates on its surface. The strong gravity field of the compact white dwarf compresses the increasingly thick layer of hydrogen gas until a thermonuclear chain reaction takes place. This is a nova explosion. After the nova explosion, the whole process starts afresh at the beginning. Some "recurring" novae explode every couple of decades; with others, there is a much longer time between successive explosions. Several dozen novae explosions are estimated to take place in the Milky Way every year, some of which can be seen with the naked eye.

▶ Light echoes surround the explosive star V838 Monocerotis, which underwent a massive outburst in 2002.

PASSPORT

Name: Scorpius X-1
V818 Sco

Constellation: Scorpius

Sky position:
R.A. 16ʰ 19ᵐ 55ˢ
Dec. -15° 38.4'

Star chart: 12

Distance:
9,000 light-years

Brightness (optical):
12.2ᵐ

Binary period: 18.9ʰ

Masses: 0.4/1.4 x Sun

Luminosity (X rays):
60,000 x Sun

X-Ray Surprises

In 1962, American astronomers launched a sounding rocket with a Geiger counter on board. It had already been discovered that the Sun emits high-energy X-rays, and the astronomers wanted to know whether the Moon does the same. Instead, however, they discovered a bright source of X-rays far beyond the solar system, in the constellation Scorpius.

Scorpius X-1, as the object was called, is an X-ray binary star at a distance of around 9,000 light-years from the Earth. One of the two components is a relatively normal star, with about half the mass of the Sun. The other, however, is an extremely compact neutron star, the survivor of a supernova explosion. With its strong gravitational force, the neutron star attracts material from its companion. This stellar gas becomes so unimaginably hot that it radiates X-rays.

Besides low-mass X-ray binaries like Scorpius X-1, where the donor star is comparable in mass to the Sun, there are also high-mass X-ray binary star systems. They consist of a hot, massive giant accompanied by a neutron star or a black hole. A powerful stellar wind carries material from the massive star to its compact companion.

The best-known high-mass X-ray binary is Cygnus X-1 in the constellation of Cygnus. Measurements of periodic wiggles of the visible giant star (known as HDE 226868) show that the companion has an orbital period of 5.6 days and has about fifteen times the mass of the Sun. It is almost without doubt a black hole. The observed X-rays are the death throes of gas that is being swallowed up by the black hole.

▲ Jets of energetic particles are blown into space in two opposite directions along the rotation axis of the black hole.

▼ Hot gas accumulates in a rapidly spinning, X-ray–emitting accretion disk before it plunges into the black hole.

▼ Radio (purple) and X-ray (blue) observations of Circinus X-1 show that at less than 4,600 years old, it is the youngest known X-ray binary.

PASSPORT

Name: Cygnus X-1
HDE 226868

Constellation: Cygnus

Sky position:
R.A. 19^h 58^m 22^s
Dec. +35° 12.1'

Star chart: 7

Distance:
6,100 light-years

Brightness (optical):
8.9^m

Binary period: 5.6^d

Masses: 30/14.8 x Sun

Luminosity (optical):
350,000 x Sun

▲ Observations of an X-ray binary known as GRO J1655-40 show that 30% of the gas that flows toward the black hole is blown back into space.

Planet Hunt

For centuries, astronomers have speculated about the existence of planets orbiting other stars. The discovery of protoplanetary disks around newborn stars suggested that the formation of planetary systems is a relatively "normal" process. Yet the first bona fide exoplanet was not discovered until 1995. Since then, almost 2,000 have been identified, and we know that at least half of all stars are accompanied by one or more planets.

A planet orbiting another star is usually too small and too faint to be seen. Astronomers have so far only succeeded in photographing a handful of exoplanets, but their existence can also be deduced indirectly. The gravitational pull of a massive exoplanet causes periodic "wiggles" in the position of its parent star on the sky. These reflex motions can be seen in the light of the star thanks to the Doppler effect, which also allows the mass of the planet to be calculated.

The American space telescope Kepler has discovered many exoplanets using the transit method: if we see the orbit of an exoplanet edge-on, the planet passes in front of the star during each orbit, causing minute periodic dips in the brightness of the light emitted by the star. The size of these dips reveals the diameter of the planet. If the mass has also been deduced through the Doppler effect, the density and composition of the planet can also be determined.

The European Space Agency's space telescope Gaia, launched in December 2013, is expected to discover tens of thousands of exoplanets on the basis of minute periodic shifts in the positions of stars. Exoplanets have also been found using gravitational microlensing.

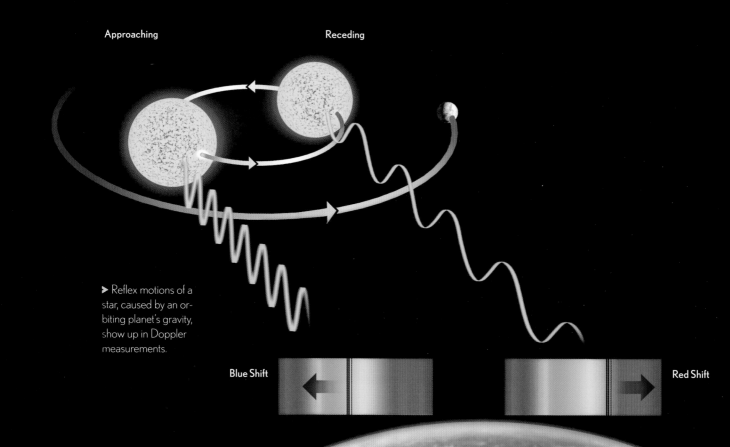

Approaching   Receding

➤ Reflex motions of a star, caused by an orbiting planet's gravity, show up in Doppler measurements.

Blue Shift   Red Shift

▼ The Kepler space telescope kept an eye on more than 150,000 stars in search of periodic brightness dips caused by planetary transits.

▲ When it transits in front of its parent star, an orbiting planet intercepts a small fraction of the star's light.

▲ The European space telescope Gaia is expected to discover tens of thousands of exoplanets through precise measurements of stellar positions.

RT
‖‖‖‖‖‖‖‖‖‖‖‖‖‖‖‖

egasi

n: Pegasus

n:

28ˢ

6.1'

2

light-years

5.5ᵐ

egasi b
)

m star:

n

od: 4.23ᵈ

Jupiter

> In July 2011, the hot Jupiter HAT-P-7b was the target of the one millionth science observation of the Hubble Space Telescope.

Hot Jupiters

Michel Mayor and Didier Queloz of the University of Geneva could not believe their eyes. In 1995, accompanying the star 51 Pegasi, which is about 50 light-years from Earth and is just visible with the naked eye, they discovered a large planet with about half the mass of the giant planet Jupiter in our own solar system. Whereas Jupiter is hundreds of millions of kilometers from the Sun and has an orbital period of almost 12 years, the new planet was less than 8 million kilometers away from the star and completed its orbit in no more than 4.2 days.

Since the discovery of 51 Pegasi b (the first exoplanet to be found orbiting a normal star), dozens of other "hot Jupiters" have been found. Some have orbital periods of less than a day. Others have been heated to such unimaginably high temperatures by their parent star that they have "swollen up" or are slowly evaporating. In many cases, they have temperatures of more than 1,000°C. In the case of some hot Jupiters, astronomers have been able to determine the composition of their atmospheres and even to measure wind speeds.

It is now clear that hot Jupiters are relatively rare; they are discovered more readily because they are easier to find than smaller planets in wider orbits. They were probably formed at much greater distances from their parent stars and later spiraled inward as a consequence of friction with the remains of the protoplanetary disk of gas and dust.

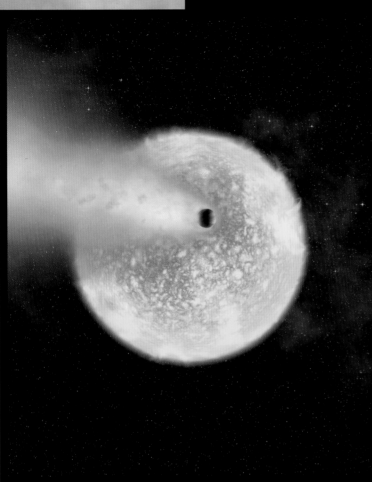

▲ HD 209458b is a giant gas planet that is so close to its parent star that it is slowly evaporating into space.

▶ Clouds of silicon dust have been discovered in the atmospheres of some hot Jupiters.

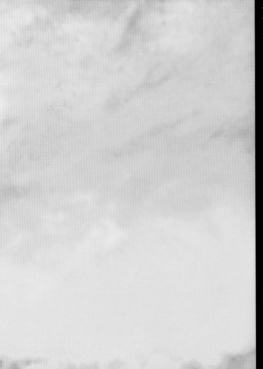

Bizarre Planets

Hot Jupiters—giant gas planets in extremely small orbits—are not the only exoplanets that defy the imagination. The hunt for planets around other stars has produced an incredible diversity of bizarre worlds, and the end is nowhere in sight. Compared with most other planetary systems in the Universe, our solar system is rather dull.

Gigantic planets have been discovered with diameters twice as large as Jupiter's and considerably larger than most brown and red dwarf stars. Others are quite compact and have enormously high densities, suggesting that they consist to a significant degree of metals or perhaps of compressed carbon (i.e., diamonds). Some planets are hardly larger or more massive than Earth, but they are so close to their parent stars that their rocky surfaces must be seas of molten lava. And there are sauna planets, with a superheated planet-wide ocean and a hot, water-vapor rich atmosphere.

The most remarkable planets are probably the small, compact worlds that trace their orbits around *pulsars*, rapidly spinning remnants of exploded stars. The first pulsar planets were discovered in 1992, when there was still no solid evidence for the existence of planets around "normal" stars. They may have been formed from material blown out into space during a supernova explosion, but no one knows for certain. The fact that there are exoplanets that even the most imaginative of science fiction writers would never dream up makes at least one thing clear: the creativity of nature knows hardly any limits.

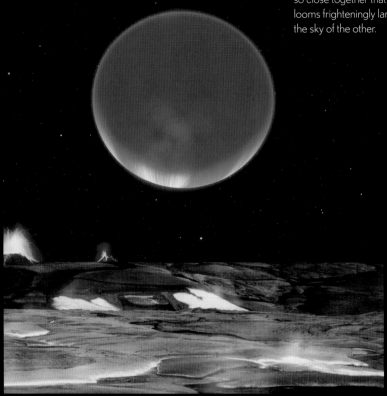

▼ Kepler-10b was the first rocky exoplanet to be discovered. It is so close to its parent star that its surface consists of molten lava.

◄ Gliese 667Cb is a super-Earth orbiting a star in a triple-star system. The stellar pair Gliese 667A and B is seen in the background.

> Kepler-186f is a planet very much like Earth, orbiting in the habitable zone of its red dwarf parent star.

Twin Sisters of Earth

How unique is Earth? That is the question that the American space telescope Kepler had to answer. By continuously keeping watch on no less than 150,000 stars for many years, Kepler found a few thousand candidate planets using the transit method. Astronomers can now finally make statistically reliable statements about the number of Earth-like planets in the Milky Way.

Small planets like the Earth are difficult to detect. They hardly cause their parent star to oscillate, and if they pass in front of it, they intercept only a small quantity of starlight. Kepler has revealed a small number of exoplanets comparable in size to Earth. In some cases, it has been determined that they have an iron core and a rocky mantle.

Such planets are more conspicuous if they orbit a red dwarf star. Because a red dwarf is much smaller than a Sun-like star, an Earth-like planet intercepts a larger percentage of the star's light during a transit. Because red dwarfs are much less massive than the Sun, they wobble more vigorously as a result of the gravitational pull of a planet orbiting around them.

The "Holy Grail" of exoplanetary research, however, remains finding a twin sister of Earth: a planet with approximately the same diameter, mass, and composition as our own world, orbiting a star similar to the Sun at a distance that is exactly right for water and life. In the Milky Way alone, there must be many billions.

◄ Three planets, including a super-Earth (in the foreground), orbit the red dwarf star Gliese 581.

> Kepler-20e (far left) and Kepler-20f (far right) are comparable in size to Venus and Earth (center).

EXOPLANETS

With a few exceptions, exoplanets have never been observed directly. If we want some idea of what these mysterious worlds look like, we have to resort to artists' impressions. They are always drawn up with great care and on the basis of the best astronomical data available. These pages show some of the distant cousins of our own Earth.

▲ Corot-7b is a rocky planet orbiting extremely close to a Sun-like star. Its surface temperature is probably around 2,000°C.

➤ Kepler-20e is a small, rocky world that is scorched by its nearby parent star.

▼ During a transit, some starlight passes through the atmosphere of exoplanet HD 209458b, allowing astronomers on Earth to study its composition.

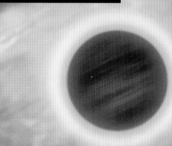

▼ The young giant planet Beta Pictoris b has been shown to have a spin period of only about 8 hours.

▲ Neptune-like planets and rocky asteroids orbit the star HD 69830.

▲ HD 149026b is a giant planet that may be the blackest and hottest one known.

▼ HD 89733b is a hot Jupiter that probably displays bright polar lights as a result of the small distance to its parent star.

➤ At least three planets are orbiting the binary star Kepler-47.

Planetary Families

Planets are "social animals." If one planet is discovered orbiting a star, there is a good chance that there will be more. The gravitational pull of two or more planets in one system causes complex, compound variations in the star's velocity; and if we look at a distant planetary system more or less from the side, we can see multiple planets passing in front of the star, each with their own orbital period.

Five planets have so far been discovered orbiting the star 55 Cancri, a Sun-like star 40 light-years away in the constellation of Cancer. Four of the five are closer to their parent star than Mercury is to the Sun; the orbit of the fifth planet is about the same size as that of Venus. The innermost planet of 55 Cancri is a "super-Earth," about twice as large and eight times more massive than our home planet.

The planetary system of the star Kepler-11 is as small and compact as that of 55 Cancri. It contains six planets, all of which transit the star. No fewer than seven planets have been found circling the star Kepler-90, the outermost of which, a giant gas planet, has an orbit roughly equal to that of the Earth around the Sun.

The planetary systems that have been found to date are therefore all very compact, but that does not mean that our solar system, where the planets are relatively widely spaced, is unique. Our current technology is simply not capable of detecting such a widespread planetary system.

➤ The richest planetary system known to date, apart from our own solar system, is HD 10180, with at least seven and maybe even nine planets.

➤ Six planets orbit the Sun-like star Kepler-11. They all show transits; sometimes three planets transit the star at once.

➤ A possible water-rich super-Earth, Kepler-22b orbits in the habitable zone of its parent star.

Goldilock Planets

If a planet is only a short distance from its parent star, it is so hot that nothing can live on it. If it is too far away, it is too cold. Life as we know it can survive only in a relatively narrow "habitable zone" around a star, where the temperature is not too high and not too low, but exactly right for liquid water. The habitable zone of our own solar system extends from just inside the orbit of the Earth almost to the orbit of Mars.

Each star has its own habitable zone. In the case of small, faint dwarf stars, this zone is of course much closer than with stars like the Sun, and for hot giant stars, it is much farther away. But for each star, it is possible to work out where a planet would need to be for it to be potentially habitable.

Quite a large number of exoplanets have been found within the habitable zones of their parent stars. In many cases, they are giant gas planets, without seas and oceans; but if one of these planets has one or more moons, those moons may be able to support life. A few Earth-like exoplanets have also been discovered in the habitable zone of their parent stars, but the latter are all faint red dwarfs.

Astronomers estimate that one in five Sun-like stars is accompanied by a small, rocky planet in the habitable zone. These planets would be real twin sisters of the Earth, but the first has yet to be found.

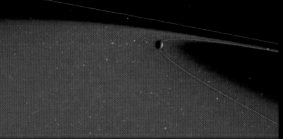

◄ Earth orbits close to the inner edge of the Sun's habitable zone.

The
DEATH
of Stars

It can take an unimaginably long time, but at some point, the life of a star will come to an end. The way it disappears from the cosmic stage is determined primarily by its mass. A small, low-mass dwarf star, for example, can last for tens or hundreds of billions of years with its hydrogen reserves because the nuclear fusion reactions in its interior take place very slowly. Eventually, the star will gradually cool off and stop generating energy.

A star like the Sun produces a somewhat more impressive spectacle. After a mere ten billion years, it will swell up to become a red giant and blow a colorful, expanding "planetary nebula" into space. After that, it will contract until it is a small, hot white dwarf that gets fainter and cooler, like a gradually dying cinder.

Massive stars are the biggest "drama queens." Their death is a cosmic spectacle that starts with a catastrophic supernova explosion visible in the farthest corners of the Universe. Even the mortal remains of these giant stars refuse to go quietly, rushing through the Universe as rapidly spinning, flashing pulsars or collapsing to become a treacherous black hole.

In the cosmos, as elsewhere, "one man's death is another man's breath." Dying stars blow material into space, from which new stars and planets are born, and that material is enriched with heavy elements created during nuclear fusion reactions. Without the death struggles of stars like the Sun and the violence of supernova explosions, a living planet like the Earth could never have evolved.

◄ A gamma-ray burst signals the death of a rapidly spinning massive star.

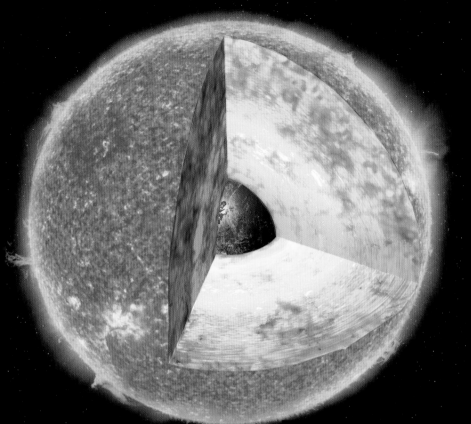

▲ In the core of a red giant, helium nuclei are fused into carbon and oxygen.

Bloated Giants

A star like the Sun produces energy through nuclear reactions in its interior. Pressure and temperature are so high there that hydrogen atoms fuse to become helium atoms. No matter how large a star is, however, the hydrogen reserves in its core are not infinite. More and more helium piles up, and eventually the fusion reactions stop. Once this happens, the star starts to collapse under its own weight. Its density and temperature increase and, in a shell around the helium core, hydrogen fusion starts to occur again.

The combustion in the shell causes the star to swell to colossal proportions, ten times larger than the Sun. Although it produces a lot of energy, this energy is radiated into space from a much larger surface area, so that the surface temperature of the star is not much higher than three or four thousand degrees. The result is a *red giant*: a large, bright but relatively cool star.

In the core of the star, however, the temperature rises enormously and, when it reaches 100 million degrees, helium atoms start fusing to form carbon. With relatively low-mass stars like the Sun, this helium fusion occurs suddenly, causing a high-energy "helium flash"; in the case of more massive stars, this takes place more gradually. Eventually, the whole process may repeat itself: the helium supply in the core becomes exhausted, the star contracts, pressure and temperature rise, and helium starts to burn in a shell, resulting in a new red giant phase.

▲ A red giantstars looms large in the sky of its scorched planet.

➤ Stars to scale: our Sun is puny compared with blue giants, red giants, and red supergiants.

Antares

Rigel

Sun

Arcturus

Sirius

Betelgeuse

Monstrous Betelgeuse

Red giants are large, but red supergiants go a step further. They are stars that were already bigger and more massive than the Sun and swell up to gigantic proportions during the helium combustion phase. The best-known example of a red supergiant is the bright star Betelgeuse, which is 640 light-years away in the constellation Orion.

Betelgeuse is more than 1,000 times larger than the Sun. If this monster were to take the place of the Sun, its surface would lie somewhere beyond the orbit of the giant planet Jupiter. And Betelgeuse is not even the biggest of the supergiants. The red giant NML Cygni, which is too far away to be seen with the naked eye, is 1,650 times larger than the Sun, with a diameter of 2.3 billion kilometers.

Betelgeuse (the name derives from the Arabic *Yad al-Jauza*, meaning "shoulder of the giant") is a relatively young star. It was born less than ten million years ago in the Orion Association, from which it is still moving away at a speed of around 30 kilometers per second. Because of its enormous mass—estimated at ten to thirty times that of the Sun—it is also evolving quickly. It has already developed a powerful stellar wind and will "soon" explode as a supernova.

The explosion may already have occurred; if Betelgeuse had exploded in the year 1600, for example, the light of the explosion would not reach Earth until 2240. When that time comes, the exploding star will be much brighter than the full Moon.

▲ Infrared observations by the European Very Large Telescope reveal the nebular gas that has been ejected by Betelgeuse.

PASSPORT

Name: Betelgeuse
Alpha Orionis

Constellation: Orion

Sky position:
R.A. 05ʰ 55ᵐ 10ˢ
Dec. +07° 24.4'

Star chart: 3

Distance:
640 light-years

Brightness: 0.4ᵐ

Mass: 10-30 x Sun

Diameter: 1,100 x Sun

Luminosity:
120,000 x Sun

Age:
8 million years

▲ Through a technique known as interferometry, astronomers have detected bright surface features on Betelgeuse.

085 | The Death of Stars

▲ Betelgeuse is the bright star in the lower left of this false-color infrared image by NASA's space telescope Wide-field Infrared Survey Explorer (WISE).

Name: Ring Nebula
M57

Constellation: Lyra

Sky position:
R.A. 18ʰ 53ᵐ 35ˢ
Dec. +33° 01.8'

Star chart: 7

Distance:
2,300 light-years

Diameter: 2.5 light-years

Age: 7,000 years

Blowing Rings in Lyra

In the 1780s, English astronomer William Herschel stumbled on a number of small, round,
nebulous spots. They reminded him of the
telescopic image of the distant planet Uranus that
he had discovered in 1781. He called them *planetary
nebulae*, a name that has endured, despite the fact
that we now know that the shells and rings of gas
have nothing to do with planets.

A *planetary nebula* is the expanding shell of gas
emitted by a star like the Sun at the end of its life.
The gas is heated up by the high-energy radiation
of the central star, in most cases a compact, hot
white dwarf, and starts to glow. The blue-green
color is due to glowing oxygen atoms, and the red
hues are produced by hydrogen and nitrogen.

The Ring Nebula in the constellation of Lyra,
discovered in 1779 by French astronomer Antoine
Darquier de Pellepoix, is one of the best-known
planetary nebulae in the sky. And with good
reason: in 1864, it was one of the first planetary
nebulae to have the composition of its light studied with a spectroscope. The measurements left no
doubt that these nebulae consist of gas heated to
extremely high temperatures.

The Ring Nebula is still expanding, at a speed
of 20 to 30 kilometers per second. From this expansion velocity, it is possible to deduce that the
nebula came into being some 1,600 years ago. In a
few thousand years' time it will hardly be visible:
the rarified gas will have cooled and dispersed into
interstellar space.

➤ The Hubble Space Telescope obtained
this detailed image of the famous Ring
Nebula in the constellation Lyra.

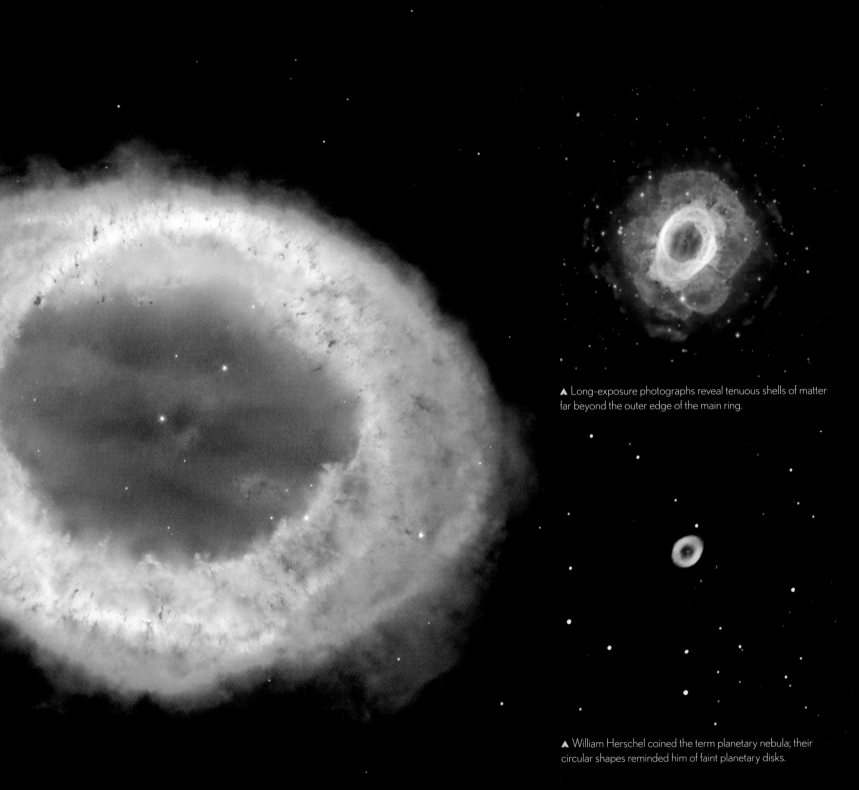

▲ Long-exposure photographs reveal tenuous shells of matter far beyond the outer edge of the main ring.

▲ William Herschel coined the term planetary nebula; their circular shapes reminded him of faint planetary disks.

PASSPORT

Name: Cat's Eye Nebula
NGC 6543

Constellation: Draco

Sky position:
R.A. 17^h 58^m 33^s
Dec. +66° 38.0'

Star chart: 1

Distance:
3,300 light-years

Diameter:
0.3 light-years
(inner part)

Age: 1,000 years

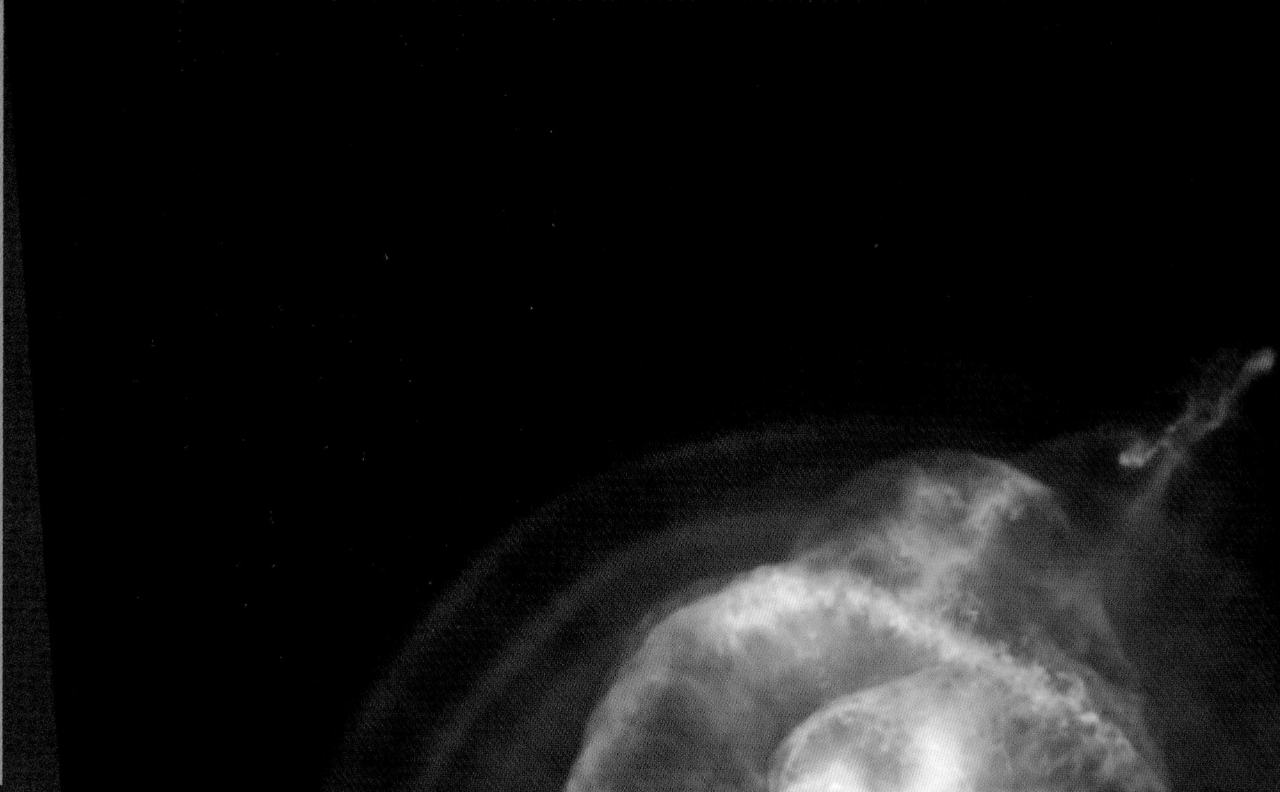

The Eye of the Cat

In 1786, high in the Northern Sky in the constellation Draco, William Herschel discovered a planetary nebula now known as the Cat's Eye Nebula. It has become particularly well known today through the magnificent detailed images made by the Hubble Space Telescope. The nebula is quite young—around 1,000 years old—and, at about 3,300 light-years from Earth, is only a small object in the night sky.

What makes the Cat's Eye Nebula so fascinating is that it is surrounded by a series of faint concentric shells of gas, which may be material blown into space by the central star in an earlier phase of its life, at intervals of a few hundred to a couple of thousand years, but astronomers are still in the dark about the precise causes. They do know, however, that the star sill has an enormously strong stellar wind, losing twenty billion tons of gas per second!

The central star is not a white dwarf, as with many other planetary nebulae, but a massive, hot Wolf-Rayet star, 10,000 times as luminous as the Sun and with a surface temperature of 80,000°C.

The Cat's Eye Nebula proved to have more surprises in store. The American space telescope Chandra discovered that there is a source of extremely high-energy X-rays at the center of the nebula. The radiation may be emitted by gas from the central star that is being captured by a companion and heated to high temperatures in the process. An invisible companion could also explain the complex structure of the nebula.

▼ The Nordic Optical Telescope at La Palma, Canary Islands, detected nebulous filaments far beyond the bright central part of the nebula.

▲ The bright X-ray source in the nebula's center is depicted blue in this false-color image.

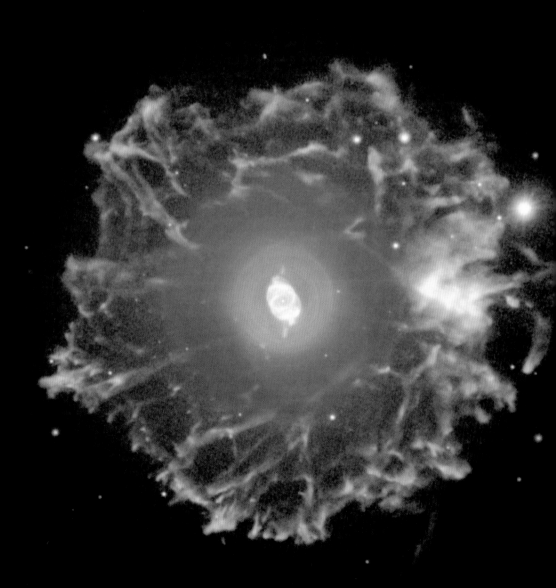

◄ The beautiful symmetry of the Cat's Eye Nebula is caught in this Hubble Space Telescope image.

▶ The European Southern Observatory's Visible and Infrared Survey Telescope for Astronomy (VISTA) in Chile obtained this detailed near-infrared view of the Helix Nebula.

▲ This is how the Helix Nebula might look when seen from the side, with a second ring almost perpendicular to the first.

Comets in the Helix

The Helix Nebula in the constellation Aquarius is the largest planetary nebula in the sky. That is primarily due to it being only about 700 light-years from the Earth. It is also relatively old, approximately 11,000 years. It is still expanding at a velocity of around 30 kilometers per second. Its diameter is currently about 2.5 light-years, more than half the distance from the Sun to the nearest star.

Detailed images of the Helix Nebula made by the Hubble Space Telescope in the 1990s showed many thousands of strange filaments of gas with a head and a tail, making them look like tadpoles. They are condensations in the gas expelled from the central parts of the nebula, roughly the same size as the solar system. Through the impact of the stellar wind from the central star, they have developed comet-like tails that point away from the center in a radial direction, creating enchanting patterns.

Similar "mini-comets" have since been discovered in many other planetary nebulae. It is not inconceivable that some of them will collapse under their own weight to form small Jupiter-like planets or cool brown dwarfs, but that is, as yet, pure speculation.

Study of the Helix Nebula has shown that it is not a simple donut-shaped gaseous nebula, as initially thought. There seems to be a second ring or shell, almost at right angles to the first. How this was formed remains unclear; perhaps the central star, which has a temperature of 120,000°C degrees, is in fact a binary, and the remarkable structure of the nebula has arisen as the consequence of a complex gravitational dance.

> ➤ Tadpole-shaped knots of gas in the Helix Nebula may look like comets, but they are much larger: almost the size of the Solar System.

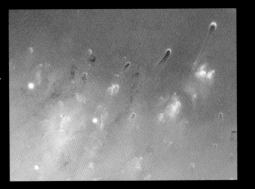

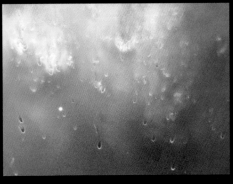

PASSPORT

Name: Helix Nebula
NGC 7293

Constellation: Aquarius

Sky position:
R.A. 22ʰ 29ᵐ 39ˢ
Dec. -20° 50.2'

Star chart: 8

Distance:
700 light-years

Diameter: 2.5 light-years

Age: 11,000 years

PLANETARY NEBULAE

As the photos on these pages show, with their amazing display of colors and their generally symmetrical circular form, planetary nebulae are among the most captivating objects in the Universe, and yet they are always associated with the death of a sun-like star and are visible for 20,000 years at most. They are the short-lived visual requiems of the cosmos.

➤ The Ant Nebula (Mz 3) in the constellation Norma.

➤ The Spiral Planetary Nebula (NGC 5189) in the constellation Musca.

➤ The Spirograph Nebula (IC 418) in the constellation Lepus.

▲ The "rectangular" planetary nebula IC 4406 in the constellation Lupus.

▲ The Little Ghost Nebula (NGC 6369) in the constellation Ophiuchus.

▲ NGC 6362 is a colorful planetary nebula in the constellation Ara.

◄ The Boomerang Nebula (IRAS 12419-5414) in the constellation Centaurus.

► The Dumbbell Nebula (M27) in the constellation Vulpecula.

◄ The Hourglass Nebula (MyCn 18) in the constellation Musca.

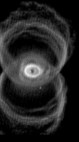

► The Eskimo Nebula (NGC 2392) in the constellation Gemini.

093

◄ The Butterfly Nebula (NGC 2346) in the constellation Monoceros.

Degenerated Dwarfs

In a few billion years, when the hydrogen supply in its core has been exhausted, our Sun will swell to become a red giant and blow its gas mantle into space in the form of a planetary nebula. What will remain will be a small, compact, hot white dwarf star, like those discovered in other planetary nebulae.

A *white dwarf* is little more than the collapsed core of a Sun-like star. That core consists mainly of carbon and oxygen atoms, but it never becomes hot enough to spark off new fusion reactions. As a result of its own gravity, however, the pressure in the carbon-oxygen core increases enormously. Atoms are in effect crushed, with their constituent nuclei and electrons compressed tightly together. This "degenerated" material has an unimaginably high density: several hundred tons per cubic centimeter.

Young white dwarfs have a surface temperature of tens of thousands of degrees. Nevertheless, they are quite small (not much larger than the Earth), so that they radiate little light: most are a few thousand times less luminous than the Sun. In the course of tens of billions of years, they slowly cool to become cold, dark cinders: black dwarfs.

The closest white dwarf is Sirius B, the companion of the bright winter star Sirius, at a distance of 8.6 light-years. Sirius B was discovered in 1862. The nearby stars Procyon and 40 Eridani are also accompanied by white dwarfs; but millions of isolated white dwarfs also roam around in the Milky Way. The closest of these, only 14 light-years away, was discovered in 1917 by the Dutch-American astronomer Adriaan van Maanen.

▲ The white dwarf star Sirius B is seen to the lower left in this Hubble Space Telescope image of the bright star Sirius.

◄ At the center of planetary nebula NGC 2440 in the constellation Puppis is an extremely hot white dwarf star.

➤ White dwarfs can be more massive than the Sun, but they are about the same size as Earth.

Out with a Bang

Stars that are much more massive than the Sun end their relatively short lives in a spectacular fashion in the form of a catastrophic supernova explosion. After the fusion of hydrogen into helium and of helium into carbon and oxygen, more nuclear reactions occur in the extremely hot interior, leading to the formation of atoms of neon, silicon, magnesium, and so forth. The energy released by all these fusion reactions offers resistance to the star's gravity.

Once stable atomic nuclei of iron are formed, the fusion reactions come to a stop. The star collapses under its own weight and with such violence that it then blows itself apart in an unimaginably powerful explosion, which is visible to the naked eye even at distances of hundreds of thousands of light-years. This supernova looks like the birth of a new star in the sky.

Less massive stars can also experience supernova explosions. A white dwarf that is part of a binary system, for example, can attract material from its companion. If it then becomes 40% more massive than the Sun, it will collapse under its own weight and explode in a supernova. The same occurs if two white dwarfs are orbiting each other, gradually move closer, and eventually collide.

A supernova flares up somewhere in the Universe on average every second. In the Milky Way, there is estimated to be a few each century. But most of them are rendered invisible by dark dust clouds.

◄ A supernova explosion blasts its way through clouds of gas that have been shed by the dying star at an earlier stage.

◄ Supernova 1994D in the galaxy NGC 4526 was almost as bright as the host galaxy's core.

095 | The Death of Stars

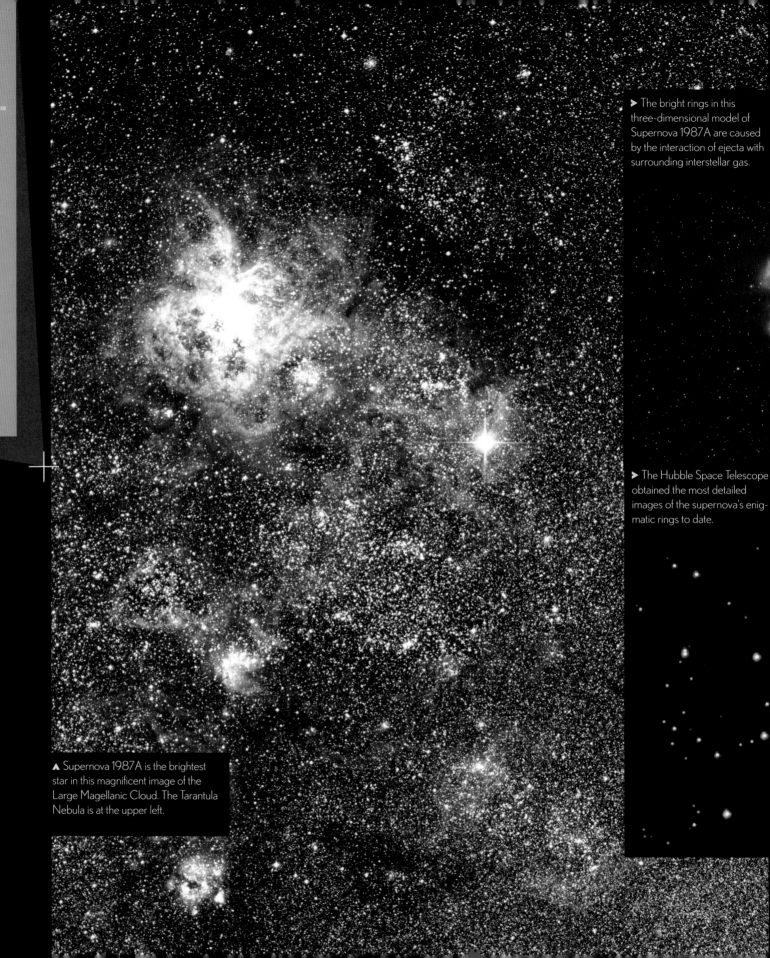

PASSPORT

Name:
Supernova 1987A

Constellation: Dorado

Sky position:
R.A. 05^h 35^m 28^s
Dec. -69° 16.2'

Star chart: 14

Distance:
168,000 light-years

Mass of star: 20 x Sun

Explosion date:
February 24, 1987

Expansion velocity:
7,000 km/s

Peak brightness: 2.9^m

▶ The bright rings in this three-dimensional model of Supernova 1987A are caused by the interaction of ejecta with surrounding interstellar gas.

▶ The Hubble Space Telescope obtained the most detailed images of the supernova's enigmatic rings to date.

▲ Supernova 1987A is the brightest star in this magnificent image of the Large Magellanic Cloud. The Tarantula Nebula is at the upper left.

Recent Violence

On February 24, 1987, while working at the Las Campanas Observatory in Chile, Ian Shelton saw a star in the sky that did not belong there. He alerted his colleagues, and it quickly became clear that it was a supernova. It was not in the Milky Way, but in the Large Magellanic Cloud, a small neighboring galaxy 168,000 light-years away.

Supernova 1987A is the most closely studied stellar explosion in history. The behavior of the "new star" was monitored for many months by telescopes on Earth and in space. Archival photos revealed the star that had "given up the ghost" during the explosion. Remarkably, the star (Sanduleak-69 202) was a blue supergiant rather than a red giant. Special detectors in underground laboratories recorded the neutrinos of the supernova: ghostly elementary particles with no electric charge and a negligibly small mass, which came directly from the collapsed core of the star.

A few months after the explosion, fluorescent gas rings were visible around the detonated star. It was gas blown into space in an earlier stage in the evolution of the star and now heated up by the energy of the supernova. Fourteen years after the explosion, the gas from the shattered star collided with these old gas rings, which became so hot that they gave off X-rays. As yet, no traces have been found of a compact remnant of the supernova, for example, in the form of a neutron star or a black hole.

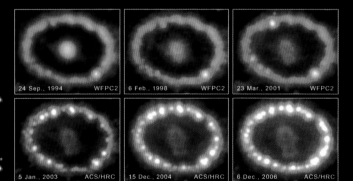

◄ In the course of two decades, numerous bright spots lit up in the gas ring surrounding Supernova 1987A.

The Death of Stars

Name: Tycho's Supernova

Constellation: Cassiopeia

Sky position:
R.A. 00^h 25^m 18^s
Dec. +64° 09.0'

Star chart: 1

Distance: 9,000 light-years

Explosion date: November 1572

Peak brightness: -4^m

Supernova type: Ia

Diameter remnant: 24 light-years

▲ The red shell in the upper left of this infrared image by NASA's space telescope Wide-field Infrared Survey Explorer (WISE) is the expanding remnant of Tycho's Supernova.

▼ Danish astronomer Tycho Brahe reported the appearance of a "Nova Stella" (labeled "I") in the constellation Cassiopeia.

A caput Cassiopeæ
B pectus Schedir.
C Cingulum
D flexura ad Ilia
E Genu
F Pes
G suprema Cathedræ
H media Chatedræ
I Noua stella.

Distantiam verò huius stellæ à fixis aliquibus in hac Cassiopeiæ constellatione, exquisito instrumento, & omnium minutorum capacj, aliquoties observaui. Inueni autem eam distare ab ea, quæ est in pectore, Schedir appellata B, 7. partibus & 55. minutis : à superiori verò

Shakespeare's Supernova?

In the opening scene of Shakespeare's *Hamlet* (written around 1600), the sentinels Barnardo and Francisco speak of a bright star "westward from the pole." According to researchers at Southwest Texas State University, the star was a bright supernova observed across Europe in early November 1572 as a "new star" in the constellation Cassiopeia. It was brighter than the planet Venus and remained visible to the naked eye for many months. Danish astronomer Tycho Brahe published an essay on it in 1573, and since then the exploding star has been known as Tycho's Supernova.

During the supernova explosion, stellar gas was expelled into space at velocities of almost 10,000 kilometers per second. An expanding shell of gas was formed, known as a *supernova remnant*. However, this leftover from Tycho's Supernova was not actually observed until 1952, with a radio telescope, and it took until 1960 to produce the first photographs. Since then, it has been imaged in detail, with the aid of infrared and X-ray telescopes in orbit around Earth.

Tycho's Supernova remnant is some 10,000 light-years away and is about 24 light-years in diameter. Astronomers have been able to determine that it was not the explosion of a massive giant star, but of a white dwarf that had exceeded its critical mass after accreting material from its companion. The companion has since also been discovered. It is racing through space at a speed of 136 kilometers per second, the direct consequence of the explosion of the white dwarf.

Kepler's Causal Connection

Thirty-two years after his mentor Tycho Brahe had seen a "new star" flare up in the constellation Cassiopeia, Johannes Kepler witnessed a supernova explosion in Ophiuchus. That was in October 1604, while Kepler was the Royal astronomer at the court of Emperor Rudolph II. A year later, he was still keeping a close eye on the fading "guest star," and 2 years after the stellar explosion, he published a book about it.

Kepler had all kinds of mystic ideas about the cosmos. He believed that the supernova explosion may have been caused (or in any case "heralded") by an exceptional conjunction of the giant planets Jupiter and Saturn a year previously in more or less the same part of the sky.

Kepler calculated that a similar conjunction of Jupiter and Saturn must also have been vis-ible in the year 7 BC. If that, too, had "produced" a supernova, it might explain the Star of Bethlehem, which, according to the gospel of St. Matthew, heralded the birth of Christ.

We know now there is no link between planetary conjunctions and supernovae. Like Tycho's Supernova in 1572, the one Kepler saw in 1604 was caused by the thermonuclear explosion of a white dwarf at about 20,000 light-years from Earth.

X-ray measurements by NASA's space telescope Chandra show that the expanding shell of gas from the supernova is asymmetric: there are more iron atoms on one side than on the other, probably because the exploding star had a nearby companion. Mass transfer from the companion eventually proved fatal for the white dwarf.

PASSPORT

Name: Kepler's Supernova

Constellation: Ophiuchus

Sky position:
R.A. 17ʰ 30ᵐ 42ˢ
Dec. -21° 29.0'

Star chart: 12

Distance: 20,000 light-years

Explosion date: October 1604

Peak brightness: -2.5ᵐ

Supernova type: Ia

Diameter remnant: 25 light-years

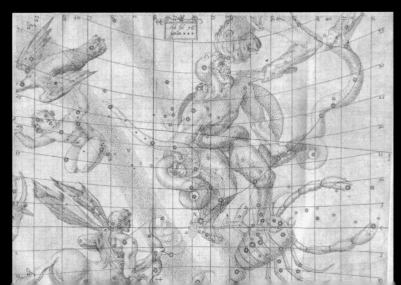

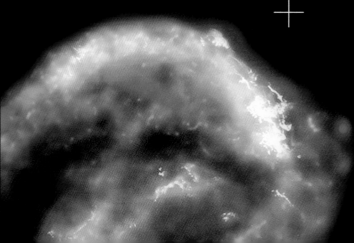

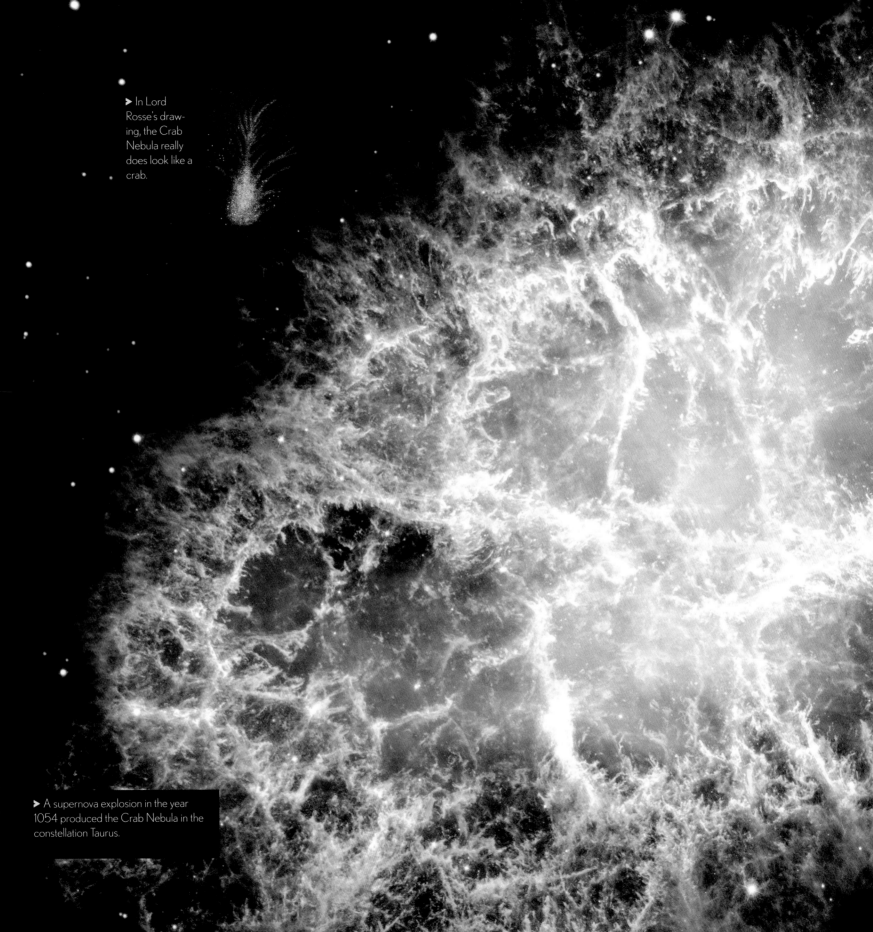

➤ In Lord Rosse's drawing, the Crab Nebula really does look like a crab.

Deep Space

100

➤ A supernova explosion in the year 1054 produced the Crab Nebula in the constellation Taurus.

An Eleventh-Century Catastrophe

A supernova explosion is at least as impressive as the appearance of a bright comet or a total eclipse of the Sun. All these strange celestial events were recorded many centuries ago by astronomers and astrologers in Egypt, China, and Babylon. Chinese chronicles from the Sung Dynasty, for example, refer to a bright "guest star" that appeared in the constellation Taurus in July 1054. For a few weeks, the new star was even visible during the day.

The supernova of 1054 was the first whose remnant was identified. The English astronomer John Bevis had discovered a small, faint nebula in Taurus as early as 1731. The nebula, later called the Crab Nebula because of its striking shape, is 6,500 light-years from Earth and is around 12 light-years in diameter.

In the early twentieth century, astronomers discovered that the Crab Nebula is expanding at a velocity of 1,500 kilometers per second. Calculating backwards, it is evident that the expansion must have started in the mid-eleventh century, around the time that the Chinese guest star appeared. It was astronomers Nicholas Mayall and Jan Hendrik Oort and sinologist Jan Duyvendak who finally proved without doubt that the Crab Nebula was indeed the expanding shell of gas from the supernova explosion of 1054.

The Chinese chroniclers witnessed the terminal explosion of a massive giant star. The mortal remains of this star were found in 1968, in the form of a tiny, supercompact neutron star that spins on its axis more than thirty times a second and emits short pulses of light, radio waves, and X-rays. It is primarily the high-energy radiation from this pulsar that makes the gas of the Crab Nebula glow.

> X-ray observations (light blue) by NASA's Chandra X-ray Observatory reveal the hot pulsar wind nebula.

PASSPORT

Name: Crab Nebula M1

Constellation: Taurus

Sky position:
R.A. 05ʰ 34ᵐ 32ˢ
Dec. +22° 00.9'

Star chart: 3

Distance:
6,500 light-years

Explosion date:
July 1054

Peak brightness: -6ᵐ

Supernova type: II

Diameter remnant:
12 light-years

Rotation period:
0.0335s (pulsar)

Eyewitnesses Sought

Not every supernova explosion in the Milky Way produces a magnificent spectacle in the night sky. More than 300 years ago, a massive star must have exploded in the constellation of Cassiopeia, but it apparently passed by unnoticed. Perhaps the explosion was concealed by thick, absorbent dust clouds, or it was an unusual sort of explosion producing little visible light.

The supernova remnant of the seventeenth-century explosion was discovered, however: in 1947, as a source of powerful radio waves (known as Cassiopeia A) and again in 1950 as an extremely faint, shell-shaped nebula. The gaseous shell has a diameter of around 10 light-years, is approximately 10,000 light-years away and is expanding at a velocity of about 5,000 kilometers per second.

Cassiopeia A has since been studied in close detail by the space telescopes Hubble (in visible light), Spitzer (at infrared wavelengths), and Chandra (in X-rays). The colors of the expanding supernova remnant reveal the composition of the gas. Green shows the presence of oxygen, red of sulfur, and blue of hydrogen and nitrogen.

The infrared space telescope Spitzer also discovered that the neutron star left behind by the explosion is still pretty active. In the middle of the last century, it underwent an explosion of high-energy radiation, heating up the surrounding dust clouds.

Did no one really see the star explode? Perhaps someone did, after all: in August 1680, English astronomer John Flamsteed saw a star that he called 3 Cassiopeiae, but it was never observed again. Who knows, it may have been the highly absorbed supernova.

➤ The star 3 Cassiopeiae, cataloged by John Flamsteed, does not appear on any map. It may have been the supernova that produced Cassiopeia A.

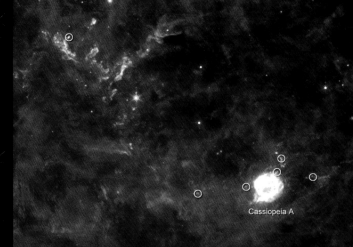

Cassiopeia A

The Death of Stars

PASSPORT

Name: Cassiopeia A

Constellation: Cassiopeia

Sky position:
R.A. 23ʰ 23ᵐ 26ˢ
Dec. +58° 48.0'

Star chart: 2

Distance: 10,000 light-years

Explosion date: 1680?

Peak brightness: 6ᵐ?

Supernova type: IIb

Diameter remnant: 10 light-years

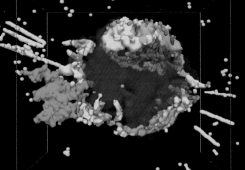

◄ Plumes of silicon and iron (yellow and green) are ejected by the supernova that created Cassiopeia A in this three-dimensional computer model.

▲ Supernova remnant Cassiopeia A has been studied at optical, infrared, and X-ray wavelengths, as shown in this false-color composite.

Hyperexplosions in the Universe

Some supernova explosions radiate as much light as the galaxy in which they are located; but some go even further than that. Gamma-ray bursts can produce as much energy in one second as the Sun in ten billion years. They are the most powerful explosions in the cosmos, and are the "death cries" of extremely massive, rapidly spinning stars that collapse to become black holes at the end of their lives.

As the name implies, gamma-ray bursts emit the largest proportion of their energy in the form of gamma rays. Fortunately, they do not penetrate the Earth's atmosphere. The mysterious bursts were therefore not discovered until 1967, by Vela military satellites searching for gamma rays from illegal Russian nuclear tests.

In 1997, Dutch astronomers Titus Galama and Paul Groot were the first to determine the distance to a gamma-ray burst. The elusive explosions proved to occur in very distant galaxies. That immediately made it clear what unimaginable quantities of energy they produce.

Besides normal ("long") gamma-ray bursts, which last from a few seconds to many minutes, there are also short bursts—sometimes no longer than a tenth of a second. They probably occur when two neutron stars collide and merge to form a black hole. Both types of burst can produce an afterglow, remaining visible with terrestrial and space telescopes for many weeks.

Astronomers do not know exactly how the bursts are caused. What is certain is that a gamma-ray burst in our cosmic backyard could destroy all life in Earth in one fell swoop.

> Gamma-ray bursts are the most powerful explosions in the Universe.

> The detonation of a massive star sends powerful jets and shock waves through space.

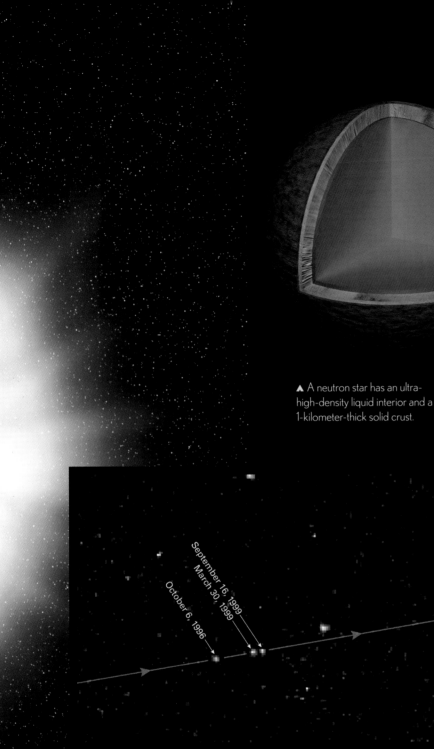

A Ball of Neutrons

▲ A neutron star has an ultra-high-density liquid interior and a 1-kilometer-thick solid crust.

September 16, 1999

March 30, 1999

October 6, 1996

▲ The isolated neutron star (and X-ray source) RX J185635-3745 is seen speeding through space in this Hubble Space Telescope composite image.

Most stars end their lives as small, compact white dwarfs; but when the most massive stars in the Universe meet their deaths through a supernova, they leave a much more extreme remnant behind. If the core of a star is more than 40% more massive than the Sun, the "degeneration pressure" of the stellar gas can no longer resist the pull of gravity. The electrons are, as it were, pushed into the atomic nuclei. The result is an unimaginably compact neutron star.

American astronomers Walter Baade and Fritz Zwicky predicted the existence of neutron stars back in 1934, but the first were not discovered until the 1960s. A little less than 2,000 have now been identified, most of which are observed as *pulsars*—rapidly flashing "radio stars." But we can still say little about their internal structure and composition with any certainty.

Neutron stars are among the most bizarre objects in the cosmos. They are more massive than the Sun, but all that material is strongly compressed into a ball with a diameter of no more than 30 kilometers. The density of a neutron star is therefore enormous: one teaspoon of neutron star matter easily weighs some five billion tons. The surface gravity is a few hundred times as strong as it is on Earth.

When a star collapses to become a neutron star, the rotation speed of the stellar core increases enormously. As a consequence, neutron stars spin on their axes tens or hundreds of times per second, and they are very hot, around a million degrees, resulting in them emitting primarily high-energy X-rays.

Cosmic Supermagnets

On March 5, 1979, an extremely high-energy gamma-ray burst swept through the solar system. Several satellites and spaces probes registered the explosion that caused it, and through triangular measurements, astronomers were able to trace it back to a supernova remnant in the Large Magellanic Cloud. Later, a handful of similar exceptional gamma-ray explosions were observed. One of these, on December 27, 2004, even had a measurable effect on Earth's atmosphere, despite being 50,000 light-years away. Astronomers believe that the explosions come from *magnetars*—neutron stars with a ludicrously strong magnetic field.

When the core of a massive star collapses, its rotation speed and the strength of its magnetic fields increase enormously. As a result of magnetohydrodynamic processes, a newborn neutron star can have magnetic fields as strong as 100 billion tesla, more than a quadrillion times as strong as the Earth's magnetic field. That is so extreme that atoms are stretched to become elongated needles. Even at a distance of 100,000 kilometers from such a cosmic supermagnet, all the information would be erased from your credit card. If you were to approach any closer than 1,000 kilometers, you would be unable to survive.

The high-energy radiation of a magnetar comes largely from the degeneration of the magnetic field. Quakes on the surface lead to incidental explosions, like the one of March 5, 1979. As a result of magnetic friction, the neutron star also starts to rotate more slowly; on average, magne-tars spin on their axes "only" once every 10 seconds. Roughly 10,000 years after they were born, little remains of the original strength of their magnetic field; there are estimated to be many millions of these "inactive" magnetars in the Milky Way.

◄ Pulsars emit beams of radiation in two opposite directions along their magnetic axis.

▲ The magnetic field of a magnetar can be more than a quadrillion times as strong as the Earth's magnetic field.

▼ Infrared observations revealed a dust ring around magnetar SGR 1900+14.

Best Clocks in the Universe

In November 1967, when Jocelyn Bell and Antony Hewish picked up short, periodic radio pulses from the Universe with intervals of 1.33 seconds, they thought at first they had found signs of alien life. The mysterious cosmic strobe light was even given the code name LGM-1, which stood for "little green men." It soon proved, however, to be a rapidly spinning neutron star. A year later, the name *pulsar*, from pulsating star, was coined.

Neutron stars have strong magnetic fields and rotate rapidly on their axes. Rotating magnetic fields generate electrical currents, and electrically charged particles spiral around magnetic field lines. The result is that beams of high-energy particles and electromagnetic radiation are blown into space along the neutron star's magnetic axis.

In most cases, as with Earth, the magnetic poles of a neutron star do not coincide with its rotational poles. This means that, as the star spins, the radiation beams sweep through space like the light from a lighthouse. If Earth lies in the path of one of these beams, we will see a short pulse during each rotation. To see a neutron star as a pulsar, we therefore have to observe it from the right perspective.

Apart from radio pulsars, there are also X-ray and gamma-ray pulsars; some also flash at visible wavelengths. A neutron star in a binary star system can sometimes be "whipped up" to enormous speeds by matter transfer from its companion. It will then become a millisecond pulsar, spinning on its axis hundreds of times per second. Millisecond pulsars rotate with astounding regularity: you can set the best atomic clocks by them.

> The energetic radiation of the pulsar in the center of the Crab Nebula is captured in this image by NASA's Chandra X-ray Observatory.

> Charged particles from a pulsar create powerful aurorae in the atmosphere of an orbiting planet.

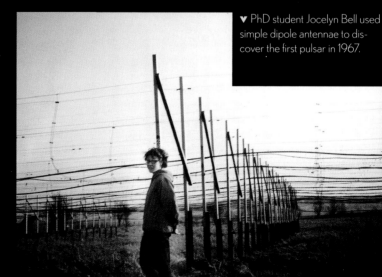

▼ PhD student Jocelyn Bell used simple dipole antennae to discover the first pulsar in 1967.

Einstein Pulsars

By far the best-known pulsar is PSR 1913+16. In 1993, it won the Nobel Prize for Physics; rather, the prize was won by Joe Taylor and Russell Hulse, who discovered in 1974 that the pulsar, with a rotation speed of seventeen revolutions per second, is part of a binary star system.

Using the 300-meter Arecibo radio telescope on Puerto Rico, Taylor and Hulse discovered that the pulses of PSR 1913+16 sometimes arrived on Earth a little closer together and at other times more widely dispersed. The pattern repeated itself every 7.75 hours. They concluded that the pulsar was accompanied by another neutron star. The two compact objects circle each other in eccentric orbits, at an average distance of a few million kilometers.

In such a bizarre binary star system, Albert Einstein's general theory of relativity starts to apply. The theory predicts that the system loses energy in the form of gravitational waves, and that the two neutron stars therefore move steadily closer to each other. And that is exactly what precision measurements of the binary pulsar showed. Every year, the rotational period decreases by 76.5 microseconds and the average distance by 3.5 meters. In around 300 million years, the two stars will collide and merge to become a black hole.

In the meantime, another binary neutron star has been discovered, PSR J0737-3039, both components of which produce observable radio pulses. This system, too, behaves just as Einstein's theory predicts. In this way, bizarre astronomical objects can be used to test theories of physics in extreme circumstances.

▼ Two rapidly spinning pulsars orbit each other as they send jets of matter and radiation into space.

▶ A binary pulsar loses energy in the form of gravitational waves—ripples in space-time that propagate with the speed of light. As a result, the two pulsars slowly spiral toward each other.

Black Holes: Prisons for Light

▲ Born in the early youth of the Universe, the young black hole in this artistic impression lacks a surrounding cloud of absorbing dust.

The more massive and compact a star is, the stronger the gravitational fields on its surface. What happens when a star is so massive that the escape velocity is higher than the speed of light (300,000 kilometers per second)? English geologist John Michel asked himself this question as early as 1783. He postulated the existence of "dark stars" with so much gravity that no light can escape.

Thanks to Einstein's theory of relativity, we know that such dark stars can indeed exist. They are known as *black holes*: black because they radiate no light themselves and holes because material in the immediate vicinity can only be sucked into them and can never escape again. According to Einstein, the speed of light is the fastest that anything can move in the natural world.

If the core of an exploded star has more than approximately three times the mass of the Sun, even the nuclear pressure of the compressed neutrons cannot offer sufficient resistance to the force of gravity. The neutron star then collapses further to become a *black hole*, a region of space with such a strong gravitational field that nothing can escape from it, not even light. No one knows what goes on in the center of a black hole; our theories of physics are not sophisticated enough to explain what happens in these black prisons for light.

"Stellar" black holes are sometimes indirectly visible if they are part of a binary star system. They then accrete matter from their companion. That gas ends up in a flat, rotating accretion disk before it disappears into the black hole and becomes so

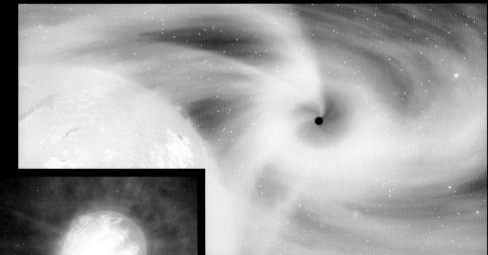

The
MILKY
WAY Galaxy

Earth is one of the eight planets in orbit around the Sun, but the Sun is not unique; together with a few hundred billion other stars, it is part of the Milky Way. You can see the Milky Way as a gigantic cosmic metropolis with a few hundred billion inhabitants—the stars. Some of these inhabitants lead a solitary life, but at least half of them have a small family of one or more planets.

The Milky Way is actually quite like a city in its structure, too. The young families with newborn children live in the quiet suburbs; closer to the center are mostly older residents, and few new stars are born there. But the city center is a hive of activity, as if a wild party is in full swing on the main square. Here one can meet many bizarre, eccentric fellow residents, like neutron stars, pulsars, and magnetars, and come across traces of earlier explosions of violence, in the form of expanding supernova remnants. A supermassive black hole sucks gas clouds and stars into its interior; in the immediate vicinity, high-energy jets of particles and radiation are blown into space.

That tumultuous center is at a safe distance of around 27,000 light-years from Earth—lucky for us, given the violent scenes that play out there. In fact, without special equipment, we would be unable to see any of it; dust clouds absorb much of the light, and we are unable to see little farther than our own sleepy suburb.

◄ The Milky Way arches over one of the smaller telescopes at the European Paranal Observatory in northern Chile.

The Milky Way in the Sky

On a dark, moonless night, it is possible to see a broad, hazy band of milky light in the sky. Sometimes it is at right angles to the horizon and runs directly above your head. At other times, it lies much flatter and is less obvious. This faint band of light surrounds Earth like a belt; one half is always above, and the other half below, the horizon.

Some parts of the Milky Way, as the band of light is called, are more conspicuous than others. The brightest regions are found toward the constellations Sagittarius, Scorpius, Cassiopeia, Cygnus, and Aquila. These bright Milky Way clouds are dissected by elongated, dark filaments. South American Indians and Australian Aboriginals recognized the shapes of animals in these "dark constellations," for example, a toad, a snake, a llama, or an emu.

In the early seventeenth century, using his home-built telescope, Galileo Galilei discovered that the hazy band we call the Milky Way is actually the combined light of countless faint stars, which are too small and too far away to be seen individually with the naked eye. The brightest parts of the Milky Way are extended "star clouds"; later, it was discovered that the darker areas are elongated dust clouds that obstruct the light of more distant stars.

If you have a telescope, you can never tire of looking at the Milky Way. Everywhere, you can see star clusters and gaseous nebulae, "nurseries" for new stars. But you need to be in a dark place, far away from towns and cities, where your view is not contaminated by artificial lights.

▼ A mosaic of images captures the full extent of the Milky Way, with its bright center and dark dust clouds.

◀ This amateur photograph, shot in Nevada, gives a good impression of what the Milky Way looks like seen from a dark location.

◀ French amateur astronomer Jacques Vincent sketched the summer Milky Way in 2012. The bright star at the upper left is Deneb.

Mapping the Milky Way

▲ Dutch astronomer Jacobus Kapteyn believed that the Sun is close to the center of a relatively small Milky Way galaxy.

William Herschel was the first to understand, at the end of the eighteenth century, that the Milky Way is a flattened conglomeration of stars, of which our own Sun is one. He even deduced the shape of the Milky Way on the basis of accurate star counts. Others later conducted much more precise counts, including Dutch astronomer Jacobus Kapteyn in the early twentieth century. We know now that their results were not entirely reliable: they took no account of the absorption of light by interstellar dust clouds, about which nothing was known at the time.

In the 1920s, it became clear that the Milky Way was just one of countless galaxies in the Universe. Many of these other galaxies have a beautiful spiral structure, but it was difficult to see whether that was also true of the Milky Way, for the simple reason that we cannot observe our own galaxy from the outside. It is like having to draw a map of a city from a street corner without leaving that street corner.

It became possible in the 1950s, thanks to the advent of radio astronomy. Radio telescopes can be used to determine the positions of interstellar clouds of cool hydrogen gas. These first radio maps, obtained by Dutch astronomers, showed conclusively that the Milky Way is a large, flattened spiral galaxy, with at least four arms and approximately 100,000 light-years across. The Sun is located somewhere in a "suburb," about 27,000 light-years from the center.

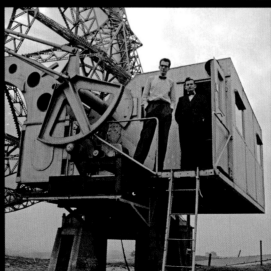

➤ Astronomers Maarten Schmidt (left) and Gart Westerhout pose at the radio dish in Kootwijk, the Netherlands, which was used to make the first map of the Milky Way.

◄ A top view of our Milky Way galaxy shows an elongated "bar" at the center, surrounded by majestic spiral arms.

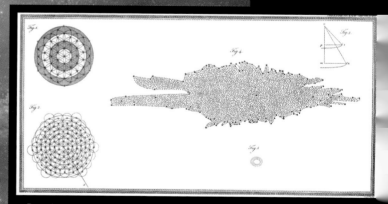

▲ Based on star counts, William Herschel constructed this model of the Milky Way galaxy.

A Peanut in a Fried Egg

Most of the spiral galaxies in the Universe are shaped like a fried egg: a thin, flat disk with a denser bulge at the center. The disk contains a lot of interstellar gas, from which new stars are still being born. The central bulge contains much less gas. New stars are rarely formed in the central bulge, and most of the stars there are relatively old. The disk of a galaxy can be compared with the maternity ward of a hospital, and the bulge is like a cosmic home for the aged.

It is difficult to see the central bulge of the Milky Way using normal telescopes, as it is largely hidden behind dark, absorbing dust clouds. But it is clearly visible using an infrared telescope, which can look right through the dust. The bulge is estimated to contain around ten billion stars in an area about 10,000 light-years in diameter.

Because we cannot see the Milky Way from the outside, it is not easy to determine the exact three-dimensional shape of the central bulge; but infrared telescopes on Earth and in space have discovered that it is elongated; from the position of the Sun, we look at the long side at an angle, which means that the Milky Way is not a normal spiral galaxy but rather a "barred" spiral galaxy.

Measurements of the positions and motion of more than twenty million red giant stars in the central bulge have also shown that the bar is shaped a little like a dumbbell. From the side, it looks like an unshelled peanut.

▼ Looming large above the dome of the European 3.6-meter telescope at La Silla, Chile, the center of the Milky Way is largely obscured by dust.

▼ Almost a million stars are visible in this infrared image of the galactic center. At visible wavelengths, they would be obscured by dust clouds.

◄ New measurements reveal that our Milky Way galaxy has a peanut-shaped heart.

▲ The plane of the galaxy is home to young stars and giant stellar nurseries like NGC 3603.

A Disk of Stars

If we could see the Milky Way from the outside, the flat disk, with a diameter of around 100,000 light-years, would be by far the most impressive sight, mainly because of its majestic spiral arms. The disk is only a few thousand light-years thick; the Sun is located at less than 100 light-years from the "central plane."

The Milky Way disk contains most of the interstellar clouds of gas and dust: cool, molecular clouds and hot, incandescent gaseous nebulae in which new stars are formed. The youngest stars are found close to the central plane of the disk. Stars can also be found a little farther above and below this central plane, but they are older and less numerous. Some astronomers therefore distinguish between the thin disk, containing the youngest stars, and the thick disk, which—like the central bulge—contains stars that were formed in the Milky Way's early youth.

The striking spiral arms are actually density waves that propagate through the slowly rotating disk of the Milky Way. Clouds of gas and dust are thereby compressed to higher densities, with the result that most star-forming regions and open star clusters are found in the spiral arms. The Sun is currently on the edge of the Orion Arm, a small spiral arm that links the large Perseus and Sagittarius Arms.

Why is the disk of the Milky Way so thin? That is because of the centrifugal force, the same force that ensures that a lump of pizza dough thrown up in the air with a rotating motion comes back down as a thin, flat disk.

◄ "Nessie," the nickname of a tendril of dust in the galactic plane, is visible in the bottom half of this infrared image. It is more than 300 light-years long.

➤ An infrared all-sky map clearly shows that most of the Milky Way's stars are concentrated in a thin central plane.

The Milky Way's Halo

Besides a flat, thin disk of mainly young stars and a central bulge of older stars, the Milky Way has a third important structural component: the halo. The halo is more or less spherical and extends far above and below the Milky Way's disk. It contains no interstellar gas clouds and therefore no longer gives birth to new stars.

For that reason, the halo contains mainly very old stars, with ages in excess of 10 billion years. They are by definition low-mass stars that are yellow, orange, or red because of their relatively low surface temperature; hotter and more massive blue-white giants live for a much shorter time. The halo also contains a few hundred *globular clusters*—spherical agglomerations of tens or hundreds of thousands of old stars.

The density of the halo gradually decreases as the distance to the center of the Milky Way increases: both the older stars and the globular clusters are more numerous closer to the center. Furthermore, they move around the center in all kinds of directions; the halo does not rotate systematically, like the disk. In 1920, American astronomer Harlow Shapley concluded, on the basis of research into the spatial distribution of globular clusters, that the Sun must be at a great distance from the center of the Milky Way.

The Milky Way's halo has no clearly defined outer limit. Ninety percent of all objects in the halo are less than 100,000 light-years from the center, but stars and globular clusters have also been discovered at distances of more than 200,000 light-years.

➤ Top and side views of the Milky Way galaxy. The halo is the spherical region that is divided in two by the galac-

▼ At a distance of 163,000 light-years, NGC 7006 is a globular cluster in the outskirts of the Milky Way's halo.

Side View

Central Bulge

Disk

Globular Clusters

Sun

Top View

Perseus Arm

Orion Arm

Sagittarius Arm

Central Bar

Central Black Hole

Norma Arm

MACHOs and Microlenses

The largest part of the Milky Way halo is found at a great distance from the Sun and Earth. Bright objects like globular clusters are easy to observe at such distances, but one needs really powerful telescopes to see faint stars. The Milky Way's halo may contain very many red, white, and brown dwarf stars, or it may be teeming with black holes. These "invisible" objects may offer an explanation for the mysterious dark matter in the Milky Way.

In 1986, Polish-American astronomer Bohdan Paczyński devised a way of tracing such "massive astrophysical compact halo objects" (MACHOs). He reasoned that if they exist, they would have to pass exactly in front of a distant "normal" star every now and then. The light of the background star would then be strengthened by the gravity of the invisible object for some weeks. A search for such microgravity lenses should thus provide information on the presence of MACHOs.

In the late 1980s and early 1990s, large-scale surveys were set up by teams in the United States, France, New Zealand, and Japan. For several years, astronomers monitored millions of stars in the Large Magellanic Cloud (a small companion galaxy to the Milky Way) to see whether their light was occasionally strengthened by the gravity of a dark object in the foreground in the halo of the Milky Way. They found a number of microgravity lenses, but not enough to conclude that all the dark matter in the Milky Way consists of MACHOs.

▼ Light from remote stars is lensed by massive astrophysical compact halo objects (MACHOs) in the galactic halo. Such MACHOs turn out to be not very numerous.

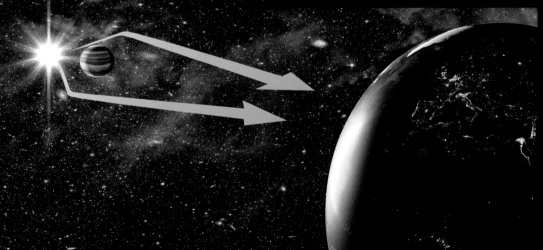

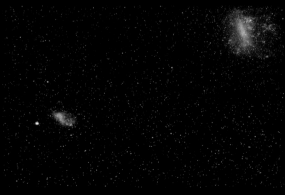

▲ Millions of stars in the Large and Small Magellanic Clouds (upper right and lower left) have been studied in the search for massive compact objects, or MACHOs.

▼ Old and dense, M15 is a globular cluster in the constellation Pegasus that harbors a black hole at its core.

▲ The European Southern Observatory's Schmidt telescope at La Silla, Chile, caught this image of globular cluster NGC 104.

Great Balls of Stars

In 1665, German amateur astronomer Johann Abraham Ihle discovered a small, faint, spherical nebula in the constellation Sagittarius. Eighty years later, eight of these round nebulae had been found, including Omega Centauri, a very bright object in the southern constellation Centaur; but their true nature remained a mystery. That did not change until 1764, when French astronomer Charles Messier saw individual stars in a small nebula that he recorded in his catalog as number M4.

More than 150 globular clusters, often referred to simply as *globulars*, have now been identified in the Milky Way. Some, like Omega Centauri, 47 Tucanae, and M13 in the constellation Hercules, are visible with the naked eye. Most of them, however, require use a telescope, and only powerful instruments can see individual stars.

Most globular clusters contain many tens or hundreds of thousands of stars. In the center, the stars are packed very densely together. The night sky must look spectacular from a planet around one of these stars, with many thousands of extremely bright stars. Moving outward, the density gradually decreases, but globulars have no sharply defined outer limits. In most cases, however, half of their starlight is produced within a radius of around 30 light-years.

Globular clusters can be found everywhere in the halo of the Milky Way, but they are concentrated in the direction of its center. They are by far the oldest objects in the Milky Way, and astronomers are still largely in the dark about how they originated.

➤ In the sky, 47 Tucanae looks like a fuzzy star. In reality, it is a huge globular cluster containing millions of stars.

▼ M13, or the Great Globular Cluster, in the constellation Hercules is the most famous globular cluster in the northern sky. The Hubble Space Telescope obtained this photograph of its core.

The Milky Way Galaxy

Name: Omega Centauri
NGC 5139

Constellation:
Centaurus

Sky position:
R.A. 13ʰ 26ᵐ 47ˢ
Dec. -47° 28.8′

Star chart: 11

Distance:
16,000 light-years

Diameter:
150 light-years

Mass: 4 million x Sun

Omega Centauri

Among the globular clusters in the Milky Way, Omega Centauri holds the record for the most stars. It is around 150 light-years in diameter and contains several million stars. Despite being almost 16,000 light-years away, it can be easily seen with the naked eye. Greek astronomer Ptolemy included it in his star catalog. Edmond Halley was the first to describe the nebular nature of the "star" in 1677, and in 1826 astronomers concluded that it is a globular cluster.

Using large telescopes on Earth and in space, it is possible to study the individual stars in Omega Centauri. In its center, they are less than a tenth of a light-year apart. They swarm around each other at speeds of almost 10 kilometers per second. According to some astronomers, the observed motions can be explained only by the presence of an intermediate-mass black hole in the center of Omega Centauri, several tens of thousand times more massive than the Sun. Hints of the presence of black holes have also been found at the center of other globular clusters.

Remarkably, Omega Centauri is not really spherical, but slightly flattened. Studies of the composition of the individual stars also indicate that there are different stellar "populations." The cluster may be the core of a dwarf galaxy that was swallowed up and ripped apart by the Milky Way. The chemical composition of Kapteyn's Star, a red dwarf only 13 light-years from the Sun, suggests that it also originated in this dwarf galaxy.

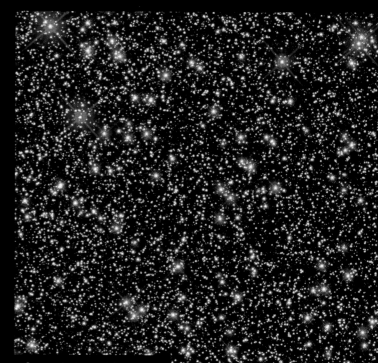

▲ Stars at the heart of Omega Centauri display a wide variety of colors on this Hubble image, one of the first obtained with the Wide-Field Camera 3 (WFC3).

➤ Slightly flattened, Omega Centauri may be the remnant of a dwarf galaxy that has fallen into the Milky Way's halo.

Stellar Streams

➤ Palomar 12 is a small globular cluster in the Milky Way that originally belonged to the Sagittarius Dwarf galaxy.

The bright star Arcturus, in the constellation Bootes, is moving through the Milky Way at high speed: 122 kilometers per second with respect to the solar system. Because Arcturus is relatively close, only 37 light-years away, it displays a significant "proper motion" through the sky of 2.3 arcseconds per year, as Edmond Halley discovered in 1718. In the twentieth century, many other stars were discovered that have a comparable proper motion and chemical composition. They are referred to collectively as the *Arcturus Stream* and are probably remnants of a small dwarf galaxy swallowed up and ripped apart by the Milky Way.

Fifteen of these stellar streams have since been identified. They can be tens or sometimes even hundreds of thousands of light-years long and may contain many millions of stars. One of the best-known is the Sagittarius Stream, the remains of the Sagittarius Dwarf Elliptical Galaxy. By far the longest stream in the Milky Way is the Helmi Stream, named after Argentine-Dutch astronomer Amina Helmi, who discovered it in 1999. The Helmi Stream appears to be wrapped around the center of the Milky Way a couple of times.

Studying stellar streams offers astronomers a unique opportunity to learn more about the origins of the Milky Way. It is suspected that, over billions of years, the Milky Way has swallowed up a large number of dwarf galaxies. By mapping the resulting stellar streams, astronomers are actually practicing a kind of cosmic archeology. The European space telescope Gaia, which was launched in December 2013, is expected to discover many new stellar streams and thus shed new light on the evolution of the Milky Way.

▲ Arcturus is the bright star in the knee of Bootes, the Herdsman. It was born in a dwarf galaxy that has since been ripped apart by the Milky Way's tidal forces.

▼ When a small galaxy is ripped apart by the tidal forces of the Milky Way, its constituent stars end up in an elongated stellar stream.

Smith's Cloud

Gail Smith was 24 in 1963 when she discovered a gas cloud on a collision course with the Milky Way. Smith, an American student working for astronomer Jan Oort in Leiden, made the discovery using the 25-meter radio telescope at Dwingeloo, in the Netherlands, at the time one of the largest in the world. After completing her studies, Smith married a Dutch doctor and exchanged astronomy for motherhood, but 45 years later "her" cloud was once again in the news.

New observations, conducted in 2008 with the American Green Bank Telescope, showed that Smith's Cloud was racing toward us at a velocity of 240 kilometers per second. Although it is still some 40,000 light-years away, the cloud has already been stretched out, by the tidal forces of the Milky Way, into an elongated shape 11,000 light-years long and 2,500 light-years wide. It will collide with the Milky Way in about 30 million years, and the resulting shock waves will probably lead to the formation of hundreds of thousands of stars.

Several more of these "high-velocity clouds" have been discovered. They consist of cool hydrogen gas that can be only seen by radio telescopes. In most cases, however, astronomers do not know how massive the clouds are, how far away they are, and from where they originally come.

Smith's Cloud is probably one of the countless gas clouds that together have formed the Milky Way over billions of years; that means our galaxy is still growing. Other high-velocity clouds were perhaps blown out of the Milky Way's disk in the aftermath of high-energy supernova explosions and are now falling back under the influence of gravity.

➤ This sketch shows the location of Smith's Cloud with respect to our Milky Way galaxy and the Sun (yellow dot).

▼ Radio observations reveal the elongated shape of Smith's Cloud, which is stretched by tidal forces of the Milky Way.

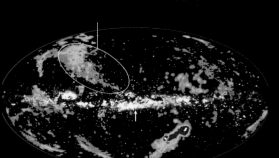

➤ Using radio telescopes, astronomers have discovered quite a large number of high-velocity clouds in our galactic neighborhood.

Stellar Gatherings

Close to the center of the Milky Way are two super-compact, young star clusters. They cannot be seen using a normal telescope, as their light is obscured by thick dust clouds in the Milky Way's disk. But infrared telescopes have detected the heat radiation of the individual stars, and the two clusters have also been observed using radio and X-ray telescopes.

The Arches Cluster is just over three million years old. It contains 150 massive bright stars in an area only 2 light-years in diameter. It is the densest star cluster yet discovered. The Quintuplet Cluster, so-called because it contains five bright infrared sources, is probably a little older, around 4 million years. This cluster also contains giant stars that may be as much as 100 times more massive than the Sun.

One of those stars is the Pistol Star, which radiates 1.6 million times more energy than the Sun. Despite being some 27,000 light-years away, it would be visible with the naked eye from the Earth if its light were not absorbed by interstellar dust clouds. The Pistol Star is expected to explode as a supernova within a few hundred thousand years.

Astronomers are a little confused about the Arches and Quintuplet Clusters. The two dense clusters are just over 100 light-years from the center of the Milky Way. No one understands how new stars can be born in this tumultuous environment. Perhaps the two clusters formed farther from the center and moved inward at a later stage.

▲ Seen in close-up, the Arches Cluster would look something like this. This artistic impression is based on scientific data.

▼ Despite its distance of around 27,000 light-years, the young, massive Arches Cluster has been studied in detail by Europe's Very Large Telescope.

◄ Ten million times as luminous as the Sun, the star in the core of the distant Pistol Nebula is one of the brightest known.

PASSPORT

Name: Sagittarius A*

Constellation:
Sagittarius

Sky position:
R.A. 17^h 45^m 40^s
Dec. -29° 00.5'

Star chart: 12

Distance:
27,000 light-years

Diameter: 12 billion km
(event horizon)

Mass: 4.2 million x Sun

▲ Future radio observations may reveal the "shadow" of Sagittarius A*, as shown in this computer simulation.

▲ Matter that falls into the galaxy's central black hole swirls around in a hot accretion disk before taking the final plunge.

Milky Way Monster

▲ Precise measurements of the motions of stars in its vicinity have revealed the mass of the Milky Way's supermassive black hole.

A gluttonous monster is living in the heart of the Milky Way. Mostly, it is quiet, but now and again, it will roar, devour what it can, and vomit. The monster is called Sagittarius A*. It is a gigantic black hole, four million times more massive than the Sun. With its gravity, it swallows tenuous gas clouds and whole stars. They disappear from the cosmic stage forever; anything that falls over the edge of a black hole can never return.

A *black hole* is invisible by definition. Its gravity is so strong that even light cannot escape, but black holes disturb their surroundings, and those effects can be seen. With their gravity, they "whip up" stars in the vicinity to unimaginable speeds. The elongated orbits of these whirling stars can be mapped using large infrared telescopes, and the mass of the black hole can be calculated on the basis of the velocities at which the stars move.

Practically all galaxies in the Universe conceal a supermassive black hole at their core. Sagittarius A* is the most closely studied of these monsters. As it is "only" 27,000 light-years away (about a quarter of a trillion kilometers), astronomers have been able to examine it in detail.

The "edge" of a black hole is also known as the *event horizon*. For Sagittarius A*, the event horizon lies at a radius of around six billion kilometers. A future worldwide network of radio telescopes, known as the Event Horizon Telescope, aims to image the horizon as a circular "shadow" that shows up dark against a background of glowing gas.

➤ NASA's Chandra X-ray Observatory captured the high-energy radiation from Sagittarius A* (third bright splotch from the left).

Calm Before the Storm

What is most remarkable about the black hole in the core of the Milky Way is that it is so improbably calm. Comparable supermassive black holes in other galaxies swallow up much greater quantities of gas from their surroundings. Just before the gas disappears beyond the horizon of the black hole, it is heated up so strongly that it emits enormous amounts of high-energy X-ray radiation. At the same time, most black holes blow powerful jets of radiation and electrically charged particles into space.

Sagittarius A*, by contrast, is the personification of calmness. It produces almost no X-ray radiation, and there are no signs at all of shimmering jets of particles. Yet there are many indications that the Milky Way monster is active on a moderately regular basis. The American X-ray telescope Chandra, for example, has observed "X-ray echoes" of a relatively small explosion that must have occurred around 300 years ago.

Above and below the center of the Milky Way, the Fermi Gamma-ray Space Telescope discovered giant bubbles of high-energy gamma rays around 25,000 light-years in size. These gamma rays are probably produced by the interaction of photons (particles of light) and very high-energy electrons. These electrons were blown into space a few million years ago during an extremely powerful explosion in Sagittarius A*.

The gas cloud G2, which passed the black hole at a very small distance in the spring of 2014, apparently survived and did not cause cosmic fireworks as astronomers had hoped. Sagittarius A* does however regularly produce small X-ray flares. These small "burps" probably come from comet-like objects that are swallowed up by the black hole; but when the next mega-explosion will occur, no one knows.

> Above and below the central plane of the Milky Way are huge "bubbles" of gamma rays, produced by high-energy electrons.

▲ A gas cloud known simply as G2 is ripped apart by tidal forces of the Milky Way's central black hole in this artistic impression.

> The inset shows the faint X-ray glow of hot gas close to the black hole. Cooler gas and dust glow at infrared wavelengths in this composite image.

X-RAY CLOSE-UP

Milky Way Mysteries

▲ X-ray echoes like those marked in this image reveal that Sagittarius A* must have been more active about 300 years ago.

SgrA*

In the past century, many puzzles about the Milky Way have been solved. Harlow Shapley used the distribution of globular clusters to deduce its dimensions. On the basis of statistical studies of the motions of stars, Jan Oort and Bertil Lindblad discovered that the Milky Way is rotating. Radio astronomers have mapped its spiral structure, and infrared telescopes have given us a glimpse of its center, which is concealed from sight by dust clouds. Even the black hole in the core of the Milky Way has given up some of its secrets.

Yet our "cosmic city" still holds many mysteries, to a large extent because we cannot observe it from the outside. A large part of the galaxy, from our perspective, which is "behind" the center, is impossible to observe. It is difficult to discover the exact shape and structure of the central bulge, although it is almost certainly an elongated bar, like that in other barred spiral galaxies. We know little about how the Milky Way has swallowed up other dwarf galaxies in the course of billions of years, and how the mysterious dark matter in the Milky Way is distributed is completely unclear.

The European space telescope Gaia aims to change all this in the coming years. Gaia will make extremely accurate measurements of the positions, distances, velocities, and chemical composition of no less than a billion stars. Armed with the resulting three-dimensional map these measurements will produce, astronomers can set about solving the remaining mysteries of the Milky Way.

◄ X-ray studies of Sagittarius A* reveal that it is a slow eater. Most of the surrounding gas is blown into space before it can disappear into the black hole.

Space Telescopes

▲ NASA's Orbiting Solar Observatory, launched in 1962, was the first true space telescope.

A telescope on the ground always has to look through Earth's atmosphere when it is trained on the Universe. Even in a cloud-free sky, the atmosphere contains dust and moisture; the air vibrates continually, and many kinds of radiation are absorbed by the atmosphere, so they cannot be observed from the ground. Long before the dawn of the space age, astronomers already dreamed of overcoming these obstacles by using telescopes in orbit around Earth.

Space travel has made it possible to lift cameras and telescopes above the confines of the atmosphere. The U.S. National Aeronautics and Space Administration (NASA) launched the first Orbiting Solar Observatory in 1962, followed in the mid-1970s by simple satellites for observing cosmic X-rays and gamma rays. The first "real" space telescope, the International Ultraviolet Explorer, was launched in 1978, followed in 1983 by the American-Dutch-British Infra-Red Astronomical Satellite (IRAS).

By far the most famous space telescope of all time is, of course, the Hubble Space Telescope, a joint project of NASA and the European Space Agency (ESA). Hubble, launched on April 24, 1990, has a mirror with a diameter of 2.4 meters and is equipped with a number of cameras and spectrographs.

After a technical problem with the main mirror was repaired during a spectacular servicing mission in space at the end of 1993, Hubble has made one revolutionary discovery after another. Later maintenance flights replaced existing cameras with much more sensitive instruments. Every area of astronomy has benefited from Hubble's sharp view of space. Despite past problems with its gyroscopes, the space telescope is still operating successfully.

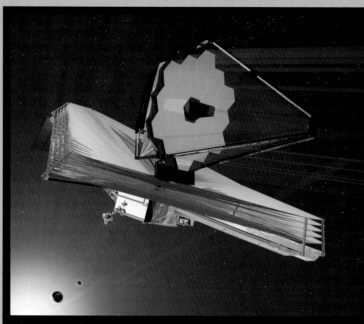

▲ Nicknamed "Son of Hubble," NASA's James Webb Space Telescope is scheduled for launch in late 2018.

▼ The European/American International Ultraviolet Explorer (IUE) operated between 1978 and 1996.

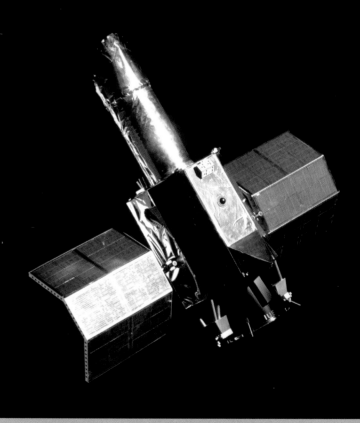

◄ In 2017, NASA will launch the Transiting Exoplanet Survey Satellite (TESS), a space telescope that will survey nearby stars for accompanying planets.

> Swift is an X-ray and gamma-ray observatory that has been designed to detect gamma-ray bursts, the most powerful explosions in the Universe.

▲ For 10 years, the ultraviolet space telescope Galaxy Evolution Explorer (GALEX) has studied the structure and evolution of galaxies.

◄ By far the most successful space telescope is Hubble, which orbits Earth at an altitude of around 600 kilometers.

131

▼ NASA's Chandra X-ray Observatory, launched in 1999, is still a productive space telescope.

▲ NASA's Wide-field Infrared Survey Explorer (WISE) focuses on the study of comets and Earth-threatening asteroids.

Deep Space

◄ Between 2009 and 2013, the European Space Agency's Herschel Space Observatory studied the distribution of interstellar molecules, including water.

◄ More than 3,500 exoplanet candidates were discovered by NASA's Kepler space telescope between 2009 and 2013.

After Hubble, NASA launched three other large space telescopes: the Compton Gamma-Ray Observatory (1991), the Chandra X-ray Observatory (1999), and the Spitzer Space Telescope (2003), which studies the Universe at infrared wavelengths. Compton remained active until the year 2000; Chandra and Spitzer are still in operation. In 1999, the ESA launched the X-ray telescope XMM-Newton (X-ray Multi-Mirror Mission), which is also still functioning.

Over the years, many smaller space telescopes have been launched, often with a specific research objective. They include the Galaxy Evolution Explorer (GALEX, ultraviolet, 2003), the Swift (gamma-ray bursts, 2004), Wide-field Infrared Survey Explorer (WISE, infrared, 2009), and Herschel (submillimeter radiation, 2009). The European Herschel Space Observatory was launched together with Planck, a space telescope designed to study the cosmic background radiation. One of NASA's most successful space telescopes is Kepler (2009), which is hunting for planets around other stars. At the end of 2013, ESA launched the space telescope Gaia, which will make precision measurements of a billion stars in the Milky Way over a 5-year period.

Currently, there are no plans for large, new X-ray telescopes in space, but NASA will be launching a colossal infrared telescope in 2018. This James Webb Space Telescope, with a mirror 6.5 meters in diameter, is considered the successor to Hubble. Two other new space telescopes, the American Transiting Exoplanet Survey Satellite (TESS, 2017) and the European Planetary Transits and Oscillations of stars (PLATO, 2024), will search for exoplanets. An idea for the very distant future is to build a large telescope on the dark side of the Moon.

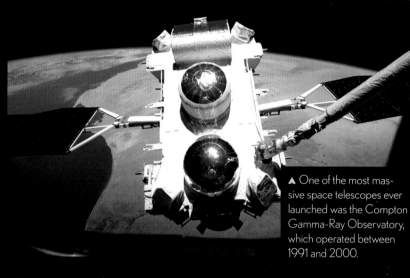

▲ One of the most massive space telescopes ever launched was the Compton Gamma-Ray Observatory, which operated between 1991 and 2000.

▶ The European Space Agency's Planck Surveyor, named after German physicist Max Planck, revealed details of the Big Bang, which spawned our Universe.

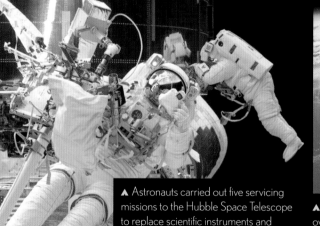

▲ Astronauts carried out five servicing missions to the Hubble Space Telescope to replace scientific instruments and broken parts.

▲ The European Space Agency has its own X-ray facility in space, called XMM-Newton.

▼ The Infrared Astronomical Satellite (IRAS) was the first space telescope to map the full sky at infrared wavelengths.

Deep Space

▲ Some four billion years from now, the Andromeda Galaxy is expected to collide with our own Milky Way.

The
LOCAL
Group

The Milky Way is not the only galaxy in the Universe. Around 100 years ago, it was established conclusively that many small, faint nebulae in the night sky are in fact distant galaxies, far away from the Milky Way.

All these galaxies are not evenly distributed across the Universe. Just as cities on Earth are often concentrated in larger agglomerations, galaxies also form both smaller and larger groups, and just as cities are often surrounded by smaller suburbs and villages, larger galaxies are accompanied by smaller "dwarf" galaxies.

Two of the companions of the Milky Way are so close to us that they can easily be seen with the naked eye, as large cloudlike nebulae. Because they can be seen only from the southern hemisphere, they were not described until a European, the Portuguese explorer Ferdinand Magellan, described them in the early sixteenth century. Since then, they have been known as the Magellanic Clouds.

Besides the two Magellanic Clouds, the Milky Way is accompanied by another two dozen small, faint dwarf galaxies. At a much greater distance is the Andromeda Galaxy, the Milky Way's nearest large neighbor. This spiral galaxy is also surrounded by two relatively large galaxies and a sizable number of small satellite galaxies.

Together with the Andromeda Galaxy, the slightly more distant Triangulum Galaxy and a number of smaller galaxies, the Milky Way makes up the *Local Group*. If you also count all the dwarf galaxies, the Local Group has more than fifty members.

Satellite Galaxies

▼ The Fornax dwarf galaxy, a satellite of our Milky Way, is little more than a huge swarm of stars. Dwarf galaxies contain almost no gas.

The Local Group has three large spiral galaxies (the Milky Way, the Andromeda Galaxy, and the Triangulum Galaxy) and about ten smaller elliptical or irregularly shaped galaxies. By far the most numerous are the *satellite galaxies*, which swarm around the large spirals like mosquitoes around a lamp.

Remarkably, no satellite galaxies have been discovered around the Triangulum Galaxy (with the possible exception of the small galaxy known as LGS3), but both the Milky Way and the Andromeda Galaxy have a large number of satellite galaxies. They are often no more than inconspicuous agglomerations of a few million stars, some thousands of light-years across.

The two large spiral galaxies probably grew by swallowing up small satellite galaxies, a process that is still going on. The Sagittarius Dwarf Galaxy, for example, has been considerably elongated by the tidal forces of the Milky Way and will eventually be absorbed by it, and the large globular cluster Omega Centauri may be the remnant core of a dwarf galaxy that was torn apart.

Research into satellite galaxies should shed more light on the origins of large spiral galaxies, about which little is known with any certainty. According to cosmological theories and computer simulations, large galaxies should have many more small satellite galaxies than have actually been observed. Perhaps those hundreds of satellites are there but consist mainly of dark matter and contain almost no stars.

The spatial distribution of satellite galaxies also presents astronomers with a puzzle. The small companions of the Andromeda Galaxy, for example, all lie more or less in one flat plane, and no one has come up with a good explanation why that should be the case.

▶ The Pisces dwarf is probably a satellite companion of the Triangulum Galaxy.

◀ M32 is a relatively small elliptical galaxy orbiting the Andromeda galaxy. In its core is a massive black hole.

Deep Space

Clouded Out

In the tenth century, Persian astronomer Abd al-Rahman al-Sufi wrote in his *Book of Fixed Stars* about small, indistinct clouds in the night sky in the southern hemisphere, but the Milky Way's two largest companions did not become known in the Western world until 1519, after Portuguese explorer Ferdinand Magellan returned from his round-the-world voyage. They are now known as the Large and the Small Magellanic Clouds.

The Large Magellanic Cloud is about 15,000 light-years in diameter and contains a few billion stars. It may once have been a small, barred spiral galaxy that has been strongly distorted by the tidal forces of the Milky Way. The cloud is about 167,000 light-years away.

The galaxy contains large quantities of interstellar gas and dust, and the level of star-forming activity is therefore much higher than in the Milky Way. The Tarantula Nebula, with the dense star cluster 30 Doradus, is one of the largest star-forming regions we have discovered; if it were only as far away as the Orion Nebula, it would never again be dark at night!

The Large Magellanic Cloud contains dozens of globular clusters and hundreds of open star clusters and planetary nebulae. Supernova 1987A also flared up in this galaxy. Moreover, it is striking that the Magellanic Clouds comprise far fewer heavy elements than the Milky Way. These "metals" (astronomical jargon for elements that are heavier than hydrogen and helium) are created over the course of time by nuclear fusion reactions in the interior of stars. Everything therefore seems to suggest that the Magellanic Clouds are considerably younger than the Milky Way.

➤ The European Southern Observatory's Schmidt telescope at La Silla, Chile, captured this stunning image of the Large Magellanic Cloud.

➤ The Large Magellanic Cloud appears to hover low above the radio dishes of the Atacama Large Millimeter/submillimeter Array (ALMA) observatory in Chile.

▼ The Tarantula Nebula is by far the largest star-forming region in the Large Magellanic Cloud and one of the largest known.

PASSPORT

Name: Large Magellanic Cloud

Constellation: Dorado/Mensa

Sky position:
R.A. 05ʰ 23ᵐ 35ˢ
Dec. -69° 45.4'

Star chart: 14

Distance: 167,000 light-years

Diameter: 14,000 light-years

Galaxy type: SB(s)m

137

The Local Group

PASSPORT

Name: Small
Magellanic Cloud

Constellation: Tucana

Sky position:
R.A. 00ʰ 52ᵐ 45ˢ
Dec. -72° 49.7'

Star chart: 14

Distance:
200,000 light-years

Diameter:
7,000 light-years

Galaxy type:
SB(s)m pec

▼ To the lower right of the Small
Magellanic Cloud is the large globular cluster 47 Tucanae, which is part of
our own Milky Way galaxy.

Leavitt's Legacy

The Small Magellanic Cloud is farther away than
its big brother and contains far fewer stars, just a
couple of hundred million. It is therefore also less
conspicuous in the sky, although on a moonless
night, it can easily be seen with the naked eye in
the southern hemisphere. Like the Large Magellanic Cloud, it may be a small, barred spiral galaxy
that has been strongly distorted by the tidal forces
of the Milky Way.

The Small Magellanic Cloud played an important role in determining the cosmic distance
scale. At the end of the nineteenth century, this
companion to the Milky Way was regularly photographed by astronomers from Harvard University, which had opened an observing station in
the southern hemisphere at Arequipa, Peru. The
photographic plates were measured and analyzed
in Cambridge, Massachusetts, by young astronomer Henrietta Leavitt.

Leavitt discovered a large number of Cepheids
in the Small Magellanic Cloud. These are stars that
get fainter and then brighter again over a period of
a few days or weeks. A relationship was proved to
exist between the pulsation period and the average
luminosity of a Cepheid: faint stars have a shorter
period than luminous ones. Once the "period-lumi-
nosity relation," also known as the *Leavitt law* (see p.
67), had been properly verified, it could be used to
determine the distance to other galaxies.

▲ The Small Magellanic Cloud contains
many star-forming regions.

▼ Henrietta Leavitt (third from left) was one of the female as-
sistants at Harvard University.

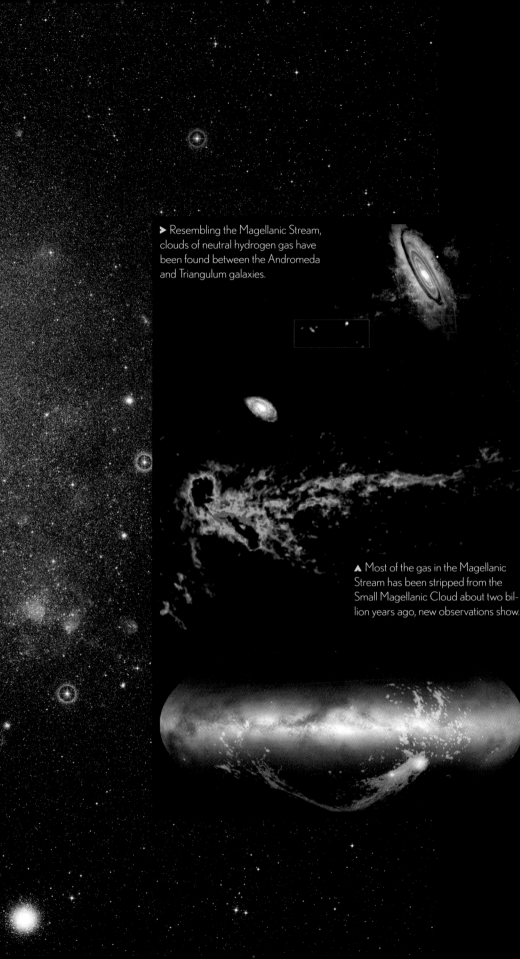

A Bridge of Gas

> Resembling the Magellanic Stream, clouds of neutral hydrogen gas have been found between the Andromeda and Triangulum galaxies.

▲ Most of the gas in the Magellanic Stream has been stripped from the Small Magellanic Cloud about two billion years ago, new observations show.

In the 1960s, astronomers discovered a "bridge" of neutral hydrogen gas between the Large and the Small Magellanic Clouds. This Magellanic Bridge, however, pales in significance compared with the much longer Magellanic Stream, which was discovered in the 1970s. The Magellanic Stream also consists of cold, neutral hydrogen gas, which means it can be observed only using radio telescopes; but it is at least 600,000 light-years long and links the two Clouds to the Milky Way.

The hydrogen gas in the Magellanic Stream was at some time in the past extracted from the two Magellanic Clouds by tidal forces. By mapping the spatial distribution of the gas, it is possible to learn more about the paths the Milky Way's two companions have taken in the past billions of years.

That analysis shows that the two Magellanic Clouds passed very close to each other around 2.5 billion years ago. Half a billion years later, the Small Magellanic Cloud had lost a substantial part of its reserves of interstellar gas, and another part of the Magellanic Stream appears to have been extracted from the Large Magellanic Cloud more recently.

The Hubble Space Telescope has made precision measurements of the movements of individual stars in the two Magellanic Clouds. The results show that both small galaxies are moving at high speeds of 300 to 400 kilometers per second, which suggests that they are not in orbit around the Milky Way, but are passersby that have been pulled out of shape by its tidal forces.

◀ The Magellanic Stream appears as a pink ribbon of gas on this radio image of the sky, obtained with radio telescopes around the world.

139

The Local Group

PASSPORT

Name:
Andromeda Galaxy
M31

Constellation:
Andromeda

Sky position:
R.A. 00^h 42^m 44^s
Dec. +41° 16.1'

Star chart: 2

Distance:
2.5 million light-years

Diameter:
120,000 light-years

Galaxy type: SA(s)b

Black hole mass:
30 million x Sun

▼ Like our own Milky Way, Andromeda is a giant spiral galaxy surrounded by smaller companions.

▶ Ultraviolet observations of Andromeda, carried out by the Galaxy Evolution Explorer (GALEX) Space Telescope, reveal distinct rings of hot, young stars.

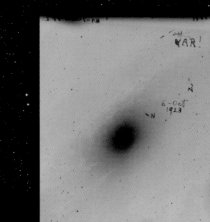

◀ Edwin Hubble was the first to discover a variable star (VAR) on a 1923 photographic plate of the Andromeda galaxy.

▼ Using the Hubble Space Telescope, astronomers can easily resolve individual stars in the disk of the Andromeda galaxy.

▼ Infrared radiation from cosmic dust (orange) and high-energy X-rays from neutron stars and black holes (blue) are combined in this composite image of the Andromeda Galaxy.

Cosmic Neighbor

On a clear, moonless night in autumn, the Andromeda Galaxy is just about visible with the naked eye for observers in the northern hemisphere, as a faint smudge of light to the northwest of the star Nu Andromedae. Around 2.5 million light-years away, it is one the most distant objects in the Universe that can be seen without using optical instruments. Like the two Magellanic Clouds, this elongated nebula was described by the Persian astronomer al-Sufi in the tenth century.

For a long time, astronomers could not agree about the true nature of spiral nebulae like the Andromeda Galaxy. Some believed them to be nebulae in the Milky Way, whereas others were convinced that they were separate "island universes" (galaxies). It was not until Edwin Hubble discovered Cepheids in the Andromeda Galaxy in

the 1920s that its distance could be determined, and it proved to be a spiral galaxy, comparable to the Milky Way.

Measurements by the Spitzer Space Telescope have shown that the Andromeda Galaxy contains about a trillion stars, almost twice as many as the Milky Way. Andromeda is also substantially larger than the Milky Way, and it is the largest galaxy in the Local Group. Strangely enough, the total mass of the Andromeda Galaxy, including dark matter, seems to be about 35% smaller than that of the Milky Way, and there also appears to be less star-forming activity. Because we cannot observe the Milky Way from the outside, and can therefore never have a complete overview of it, the Andromeda Galaxy is the best-studied galaxy in the Universe.

Dec. 17, 2010

Dec. 21, 2010

Dec. 30, 2010

Jan. 26, 2011

▲ Hubble Space Telescope close-ups of the variable star in the Andromeda galaxy, discovered in 1923 by Edwin Hubble.

The Local Group

Collision Course

In four billion years or so, the Local Group of galaxies is due to be shaken up by a cosmic "traffic accident." Around that time, the Milky Way and the Andromeda Galaxy will collide. Another two billion years later, they will merge to form a single giant elliptical galaxy. No one knows whether our own solar system will survive the collision. Perhaps the Sun and the planets will be catapulted into space by gravitational disturbances.

Astronomers have known for nearly 100 years that the two large galaxies are moving toward each other at a speed of about 110 kilometers per second; but it took them until 2012 to measure the minimal sideways motion of the Andromeda Galaxy. If it proved large enough, the two galaxies might pass each other at a short distance. This lateral velocity proved, however, to be only 17 kilometers per second, meaning that the Milky Way and the Andromeda Galaxy are indeed on a collision course.

The exact role of the Triangulum Galaxy, the third large galaxy in the Local Group, is not known. Perhaps it will collide with the Milky Way even earlier. It is also possible that, in the distant future, it will end up in orbit around the giant elliptical galaxy that will result from the collision between the Milky Way and the Andromeda Galaxy, known unofficially as *Milkomeda*.

The collision between the Andromeda Galaxy and the Milky Way will not be the first cosmic traffic accident to occur in the Local Group. Five and a half billion years ago, the Andromeda Galaxy was probably created as the result of the collision and merger of two smaller galaxies.

> The spiral galaxy M33, also known as the Triangulum Galaxy, is the smallest of the three large members of our Local Group.

▼ A few billion years from now, Earth's night sky will be ablaze with the light of the Andromeda galaxy as it merges with our own Milky Way into a giant elliptical galaxy, dubbed Milkomeda.

▼ Radio observations (purple) reveal that clouds of cold hydrogen gas extend way beyond the visible disk of the Triangulum Galaxy.

Galactic Triangulation

Like the Andromeda Galaxy, the Triangulum Galaxy is named after the constellation in which it is located, the small constellation, Triangulum, which lies between Andromeda and Aries. The faint spiral galaxy was only discovered after the invention of the telescope, but under extremely favorable circumstances, it is also visible with the naked eye, at least for people with excellent eyesight. The Triangulum Galaxy is the smallest of the three large galaxies in the Local Group. With a diameter of around 50,000 light-years, it is half the size of the Milky Way and contains about forty billion stars. Interestingly, this spiral galaxy does not appear to have a black hole in its core.

At the beginning of the twentieth century, Dutch-American astronomer Adriaan van Maanen thought he had measured the rotation of the Triangulum Galaxy. On the basis of his calculations, Van Maanen concluded that the spiral galaxy could never be more than one million light-years away. Later, however, he proved to have made a series of systematic errors of measurement. In reality, the Triangulum Galaxy is 2.5 to 3 million light-years away. Its distance to the Andromeda Galaxy is, however, much smaller, at around one million light-years. Some astronomers feel that the Triangulum Galaxy should be seen as a large companion to the Andromeda Galaxy.

There are a number of large star-forming regions in the Triangulum Galaxy. The largest is NGC 604, a magnificent nebula of incandescent hot gas around 1,500 light-years in diameter—forty times larger than the Orion Nebula. NGC 604 was discovered in 1784 by William Herschel, long before we knew the true nature of the Triangulum Galaxy.

▼ NGC 604 is a giant stellar nursery in the Triangulum Galaxy. It was already discovered in 1784, by William Herschel.

PASSPORT

Name:
Triangulum Galaxy
M33

Constellation:
Triangulum

Sky position:
R.A. 01^h 33^m 50^s
Dec. +30° 39.6'

Star chart: 2

Distance:
2.5-3 million light-years

Diameter:
50,000 light-years

Galaxy type: SA(s)cd

The Local Group

GALAXIES

If stars are the inhabitants of the cosmos, you could say that galaxies are the villages and cities in which they live. Almost all stars in the Universe are part of a galaxy; hardly any at all are to be found in intergalactic space *between* galaxies. Galaxies are therefore small islands of light in the interminable ocean of cosmic darkness.

Astronomers did not realize until the early twentieth century that our Milky Way is only one among countless galaxies in the Universe. We now know that the observable Universe contains a good hundred billion galaxies. These include majestic spiral and barred galaxies like the Milky Way, as well as colossal elliptical galaxies, misshapen "irregular" galaxies, dwarf galaxies, and starburst galaxies, which contain enormously active star-forming regions.

Galaxies are not "loners." The bigger ones are accompanied by a large number of small satellite galaxies, and they are almost always part of a group or cluster. Sometimes two galaxies pass each other at close quarters or actually collide. Such cosmic encounters generate spectacular tidal tails and give rise to "baby booms" of new stars.

Practically all galaxies have a supermassive black hole at their core. If one of these central black holes swallows up a lot of material, an enormous quantity of energy is produced in the immediate vicinity, and the galaxy can be seen from a distance of billions of light-years. These active galactic nuclei are known as *quasars*. Study of these distant galaxies has shown that the largest galaxies in the present Universe were created by collisions and mergers of countless small, irregularly shaped mini-galaxies.

◄ Thirty million light-years away in the constellation Ursa Major (the Big Dipper), NGC 2841 is a serene spiral galaxy; its spiral arms are sprinkled with star-forming regions.

Shapes and Sizes

▲ Dust clouds trace the arms of the spiral galaxy NGC 4414.

▲ The dwarf irregular galaxy I Zwicky 18 is much smaller than our Milky Way.

▲ Flat, but without spiral arms: NGC 4866 is a so-called lenticular galaxy.

▲ NGC 6217 is a beautiful example of a barred spiral galaxy.

Galaxies come in all shapes and sizes. Roughly speaking, they fall into two groups: spiral galaxies and elliptical galaxies. Spiral galaxies (type S) comprise three components: a flat, rotating disk, which contains the gas and dust-rich spiral arms and in which most star forming activity occurs; a central "bulge" of much older stars; and an extended halo of dark matter and globular clusters. Spiral galaxies are subdivided into three types (Sa, Sb, and Sc), depending on how tightly "wound" their spiral arms are: Sa galaxies have very tightly wound arms, and Sc galaxy arms are very loosely wound.

In many spiral galaxies, including our own Milky Way, the central bulge is elongated. Little is known for certain about the origins of these central "bars." Barred spiral galaxies (classifed as SBs) are also subdivided into three types: SBa, SBb, and SBc.

Elliptical galaxies (type E) are three-dimensional agglomerations consisting almost solely of stars; they contain almost no interstellar gas. The motion of the stars is less organized than in a spiral galaxy. Elliptical galaxies come in a wide variety of shapes and sizes, from spherical (E0) to very elongated (E9), from small dwarf galaxies to giant galaxies with many trillions of stars, like the galaxy M87 in the constellation of Virgo.

Lenticular galaxies (type S0) are an intermediate type between spiral and elliptical galaxies. They have a spiral disk but consist almost entirely of a flattened central bulge so that they closely resemble elliptical galaxies. The last type of galaxy are irregular galaxies (type Irr), which are mostly relatively small agglomerations of several hundred million stars with no observable structure.

➤ The giant elliptical galaxy NGC 1132 is seen against a backdrop of fainter, more distant galaxies.

Spiral Beauty

The magnificent spiral galaxy Messier 81 (M81) in the constellation of Ursa Major is also known as Bode's Galaxy, after German astronomer Johann Elert Bode, who discovered it in 1774. M81 is around 12 million light-years away, only five times as far as the Andromeda Galaxy. It is not part of the Local Group but belongs to an independent group, together with more than thirty neighboring galaxies. Because it is relatively close, M81 can easily be seen with a small telescope.

As in all spiral galaxies, most star-forming activity in M81 occurs in the spiral arms, where most interstellar gas and dust clouds are found. The high-energy radiation from young, hot stars also heats up the dust in the spiral arms, so that it

begins to emit infrared radiation. Most of the heat radiation from the galaxy therefore comes from the spiral arms.

At the center of M81 is a supermassive black hole with a mass of around 70 million suns, seventeen times more massive than the black hole in the nucleus of the Milky Way. M81's black hole is also more active than ours and may be fueled by gas that found its way to the nucleus in the aftermath of an encounter with the nearby galaxy M82, also known as the Cigar Galaxy, around half a billion years ago.

In 1993, a supernova was discovered in the outer regions of M81. The star that came to such a catastrophic end in the explosion was identified on old photographs and proved to be a red supergiant.

▼ Imaged at various infrared wavelengths (24, 70, and 160 microns), M81 reveals its star-forming regions, located in the spiral arms.

▼ Bode's Galaxy is a loosely wound spiral at a distance of about 12 million light-years.

Galaxies

Picture Perfect Pinwheel

Like Bode's Galaxy (M81), the Pinwheel Galaxy (M101) is also in the constellation of Ursa Major. It is, however, twice as far away, at a distance of 27 million light-years, so that one must use a quite powerful telescope to see it. On the other hand, M101 is a very large galaxy some 170,000 light-years across—70% larger than the Milky Way.

M101 has a rather asymmetrical structure: its bright nucleus is not at the center of the extensive disk. Furthermore, the disk, which has loosely wound spiral arms (M101 is an Sc galaxy), contains a large number of bright star-forming regions, some of which William Herschel already described in the eighteenth century. The spiral structure of the galaxy was, however, not discovered until 100 years later, by Irish astronomer William Parsons (Lord Rosse).

M101's striking asymmetrical structure is possibly due to the gravitational tidal forces of five accompanying galaxies. The closest companion, New General Catalogue (NGC) 5474, also has an asymmetrical disk.

Twice as far again as the Pinwheel Galaxy is the spiral galaxy M100, which made history in the mid-1990s when it became the first galaxy outside the Local Group in which individual Cepheids were discovered, thanks to observations by the Hubble Space Telescope. We know that the pulsation period of these variable stars is related to their real luminosity. By comparing their luminosity to their observed brightness, the distance to the galaxy can be calculated accurately. This step was an important one toward calibration of the cosmic distance scale, one of Hubble's main objectives.

▼ Hubble Space Telescope observations of M100 played a pivotal role in establishing the distance scale of the Universe.

▶ The Pinwheel Galaxy is a giant, slightly asymmetric spiral galaxy, seen almost exactly face-on.

▼ Relatively cool dust lanes in M101 light up as yellow-green filaments in this infrared image. Red denotes dust that is warmed by young, hot stars.

Name:
Whirlpool Galaxy
M51

Constellation:
Canes Venatici

Sky position:
R.A. 13ʰ 29ᵐ 53ˢ
Dec. +47° 11.7'

Star chart: 5

Distance:
23 million light-years

Diameter:
85,000 light-years

Galaxy type:
SA(s)bc pec

Black hole mass:
1 million x Sun

Down the Whirlpool

In 1845, Irish astronomer William Parsons (Lord Rosse) trained his colossal 1.8-meter telescope on a nebulous spot in the small constellation of Canes Venatici, which had already been discovered by French astronomer Charles Messier in 1773. M51, as the nebula is officially called (it is the 51st object in Messier's catalog), proved to have a remarkable whirlpool-like structure. Parsons was the first to discover the spiral nature of some nebulae.

Today M51, nicknamed the Whirlpool Galaxy, has been imaged in detail, especially by the Hubble Space Telescope. It is an impressive two-armed spiral at a distance of about 23 million light-years. Around 85,000 light-years across, it is a little smaller than the Milky Way. We see the galaxy almost directly from "above," so its spiral structure is clearly visible.

One of M51's spiral arms ends in a practically straight tail of gas, dust, and stars. At the extreme end of this arm, a smaller accompanying galaxy can be seen, which Parsons also observed. This galaxy, known as NGC 5195, is currently a little behind M51 as seen from Earth, but computer simulations show that, several millions of years ago, it must have moved through the outer regions of the Whirlpool Galaxy.

The "collision" with NGC 5195 not only affected the spiral structure of M51, but it also led to more intense star-forming activity in the larger galaxy, in both the nucleus and the spiral arms. Infrared photos of the Whirlpool Galaxy show how the dust is distributed throughout the spiral arms; a large number of small, young star clusters are visible on the images.

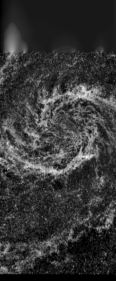

▲ The narrow dust lanes of the Whirlpool Galaxy are evident in this near-infrared image taken with the Hubble Space Telescope's Near-Infrared Camera and Multi-Object Spectrometer (NICMOS).

◄ One of the Whirlpool's spiral arms is stretched out by the gravity of a smaller companion galaxy.

▼ In 1848, William Parsons (Lord Rosse) made this delicate sketch of the spiral structure of M51.

Powerhouse Galaxy

American astronomer Carl Seyfert discovered in 1943 that the nuclei of some galaxies produce a lot of energy on a few specific wavelengths. Such emission lines suggest the presence of large quantities of extremely hot gas. Today it is widely assumed that all this activity is powered by supermassive black holes at the centers of these Seyfert galaxies, surrounded by swirling disks of gas heated to very high temperatures.

The closest Seyfert galaxy is M106, 24 million light-years away in the constellation of Canes Venatici. It was discovered in 1781 by French

▼ Like M106, M77 in the constellation Cetus is a Seyfert galaxy.

PASSPORT

Name: M106

Constellation: Canes Venatici

Sky position:
R.A. 12ʰ 18ᵐ 58ˢ
Dec. +47° 18.2'

Star chart: 5

Distance: 24 million light-years

Diameter: 140,000 light-years

Galaxy type: SAB(s)bc

Black hole mass: 40 million x Sun

Stars in a Bar

It is of course not its official name, but NGC 1300 could easily be called the Garden Sprinkler Galaxy. Its spiral arms spray out from the ends the central bar, just like spiraling water jets from a rotating garden sprinkler.

NGC 1300 was discovered in 1835 by John Herschel. It is a barred spiral galaxy about 60 million light-years away in the constellation of Eridanus, in the southern hemisphere. It is about the same size as the Milky Way and, to a certain degree, has the same structure; there is also a small bar in the nucleus of the Milky Way.

We do not know exactly how such an elongated central bar of stars is formed. It may be that gas is carried to the center from the outer regions of the

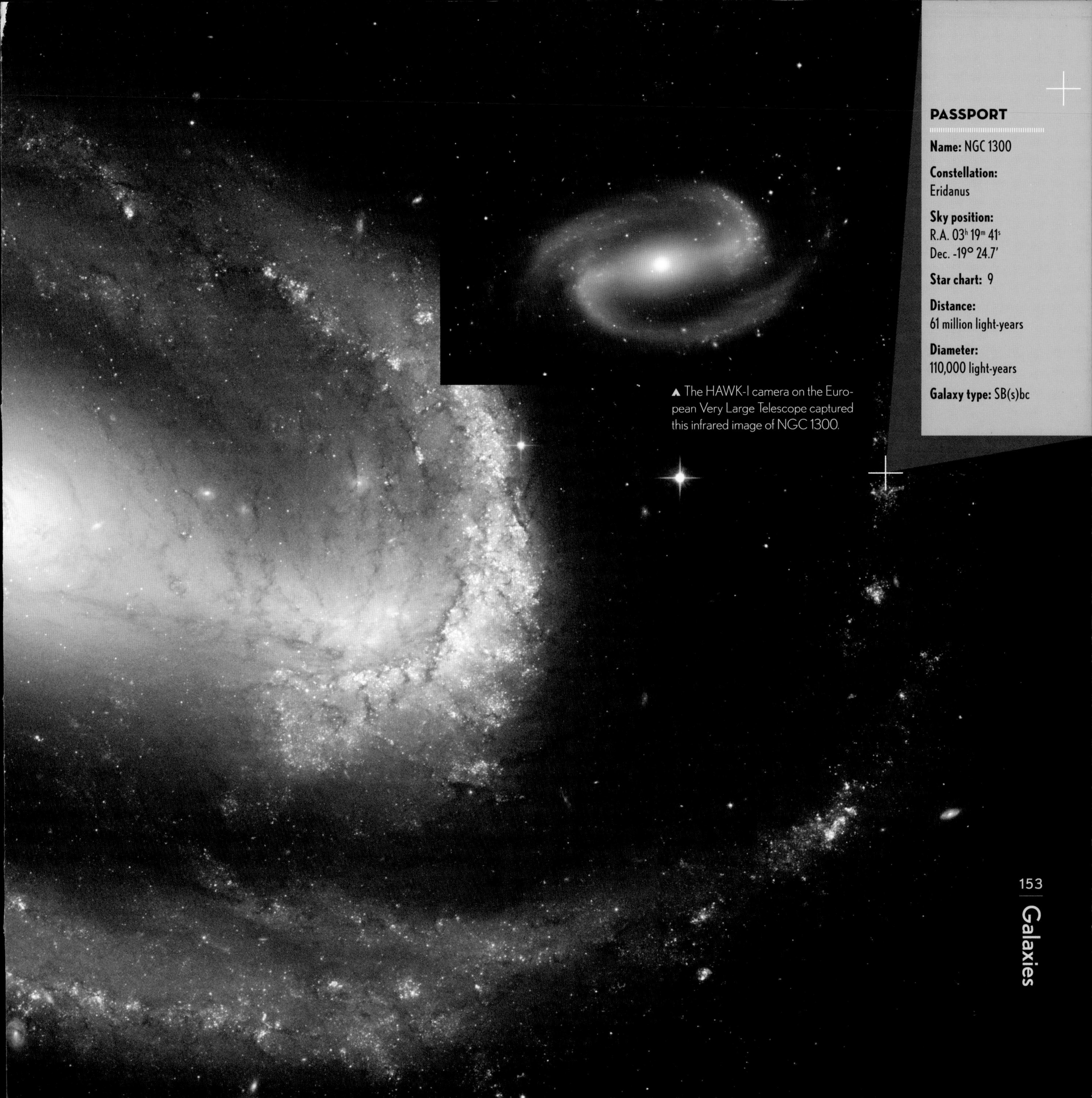

▲ The HAWK-I camera on the European Very Large Telescope captured this infrared image of NGC 1300.

PASSPORT

Name: NGC 1300

Constellation: Eridanus

Sky position:
R.A. 03^h 19^m 41^s
Dec. -19° 24.7'

Star chart: 9

Distance: 61 million light-years

Diameter: 110,000 light-years

Galaxy type: SB(s)bc

Galaxies

➤ Arp 273 is a pair of interacting galaxies. The smaller galaxy at the bottom probably passed through the larger one tens of millions of years ago.

GALAXIES

The observable Universe has some 100 billion galaxies. We can study only a small number of them in detail from the Earth; but, as the photos on these pages show, even this small selection displays an astounding diversity of forms. Our own Milky Way, the home of the Sun, would look just as impressive from the outside.

▼ No, this is not a cosmic traffic accident. In fact, NGC 3314 consists of two galaxies at different distances, one in front of the other.

▲ NGC 1097 is a Seyfert galaxy that is gravitationally interacting with a smaller companion (upper right).

◀ NGC 891 is a spiral galaxy comparable to our own Milky Way, seen exactly edge-on. The galaxy's core is just out of the image at the lower left.

▼ The spiral galaxy M83 is also known as the Southern Pinwheel Galaxy.

▼ NGC 1073 is a prototypical example of a barred spiral galaxy, 55 million light-years away in the constellation Cetus.

▼ NGC 2442 is also known as the Meathook Galaxy, because of its lopsided shape, which is probably the result of tidal forces of a past companion.

▲ In NGC 4402, gas and dust are driven out of the central plane by the pressure of hot gas in the cluster of which the galaxy is part.

◀ Because of its dark dust band, M64 is known as the Black Eye Galaxy. It is probably the remnant of a galactic merger.

PASSPORT

Name:
Antennae Galaxies
NGC 4038 / NGC 4039

Constellation: Corvus

Sky position:
R.A. 12^h 01^m 53^s
Dec. -18° 53.0'

Star chart: 11

Distance:
45 million light-years

Collision:
600 million years ago

Galaxy type:
SB(s)m pec /
SA(s)m pec

Tidal tail length:
500,000 light-years

Traffic Crash

The Hubble photo of the Antennae Galaxies is a snapshot of a cosmic "traffic accident," two galaxies colliding with each other. There is little left to see of their original regular shape (one was a "normal" spiral, the other a barred spiral galaxy), and debris is being catapulted into space around the two twisted wrecks. One could experience the tremendous devastation being wrought here only if it were possible to speed up time enormously.

The Antennae, also referred to as NGC 4038 and NGC 4039, were discovered in 1785 by William Herschel, but he thought they were a planetary nebula. His son John later discovered that there were two galaxies. They are called the Antennae because of their long, curved tails of gas and stars, created by their mutual tidal forces. The collision process has now been accurately reconstructed using computer simulations.

A little less than one billion years ago, the two galaxies approached each other and became increasingly distorted, deformed, and pulled out of shape. They actually "collided" 600 million years ago and passed through each other, just as the Milky Way and the Andromeda Galaxy will do at some point in the distant future; this event led to the formation of the "tidal tails." At present, the galaxies are again falling back toward each other, and in around 400 million years, they will merge, possibly forming a single large elliptical galaxy.

In the chaotic region between the two galaxies are colossal clouds of gas and dust, where in the future large number of new stars will form. Bright, young "super star clusters" are already visible in the Antennae Galaxies.

▲ Tidal interactions of the Antennae galaxies created the curved tails of gas and stars that give the pair their name.

➤ Computer simulations have been quite successful

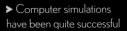

◄ Swirling dust clouds and bursts of star formation are the results of a galactic crash that occurred some 600 million years ago.

▲ Red and yellow patches mark the regions of enhanced submillimeter radiation from the Antennae, emitted by cool dust.

Cosmic Collisions

In the Universe, there are no stop signs or rules for priority. The engine of gravity is always running at full throttle, and nothing stops for oncoming traffic. Especially in relatively compact groups and clusters, galaxies regularly crash into each other. A complete *Atlas of Colliding Galaxies* has even been compiled on the basis of photos made with the Hubble Space Telescope.

A perfect example of two colliding galaxies is NGC 4676, also known as The Mice, because of their long tidal tails. The Mice are at a distance of some 290 million light-years from the Earth, in the Coma Cluster of galaxies. They must have first collided around 160 million years ago.

When two galaxies approach each other, they experience a stronger mutual gravitational pull on one side than on the other. As a result of that difference (known as tidal action), the galaxies become distorted and elongated, and curved tails of gas and stars are torn loose and catapulted outward. The practically straight "mouse tail" we see on the photo is also curved in reality, but we cannot see that clearly because we are looking at it edge-on.

The evolution of The Mice can be accurately replicated and explained using computer simulations. These also show where interstellar gas and dust is being swept together. In those regions, new star clusters are forming, which can be clearly seen in the photo as bright blue spots. Like the Antennae Galaxies, The Mice will also merge in the future to form a single large, elliptical galaxy; but that could take another few hundred million years.

➤ The Hubble Space Telescope captured this dramatic image of The Mice, two interacting galaxies with long tidal tails.

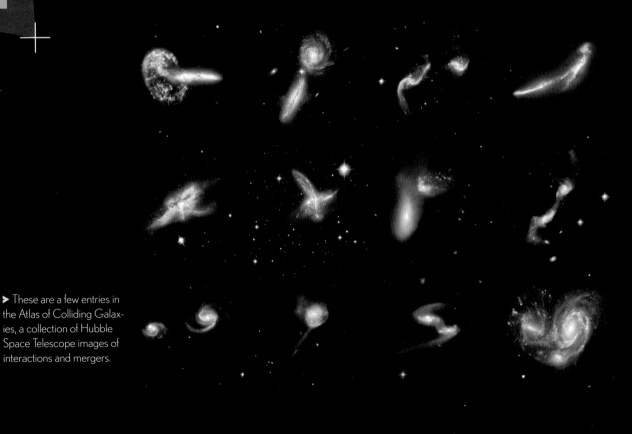

➤ These are a few entries in the Atlas of Colliding Galaxies, a collection of Hubble Space Telescope images of interactions and mergers.

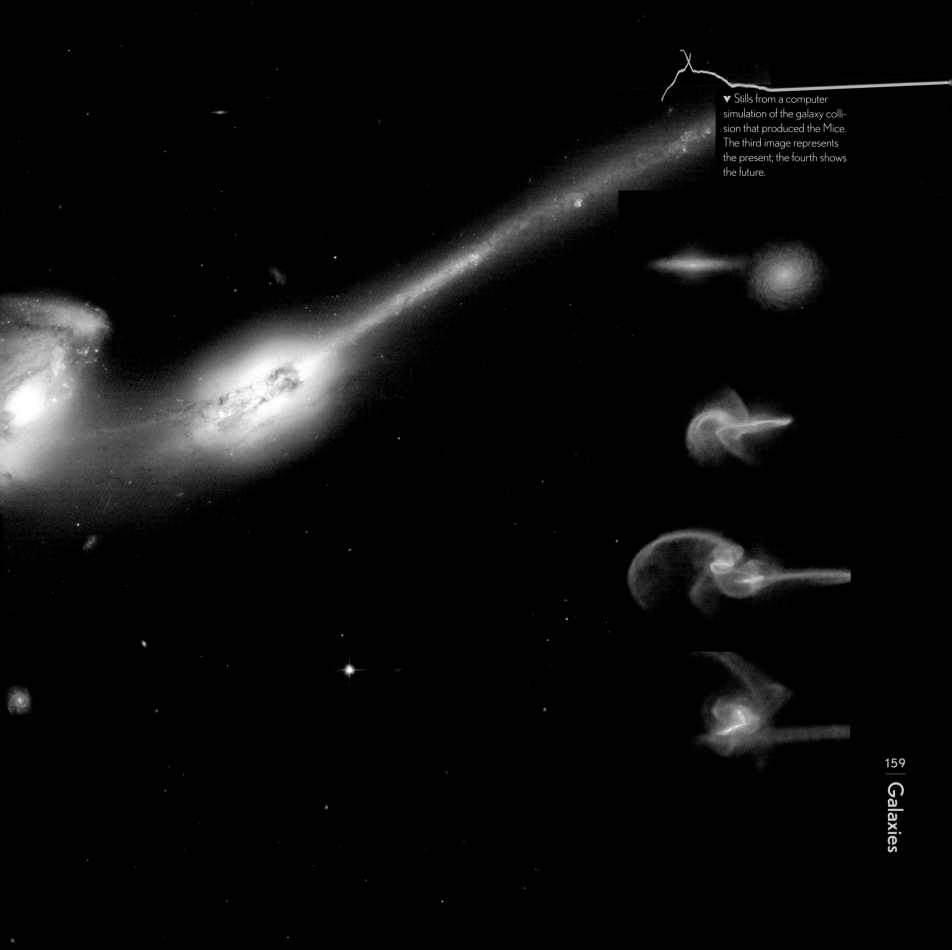

▼ Stills from a computer simulation of the galaxy collision that produced the Mice. The third image represents the present; the fourth shows the future.

Name: Cigar Galaxy
M82

Constellation:
Ursa Major

Sky position:
R.A. 09ʰ 55ᵐ 52ˢ
Dec. +69° 40.8′

Star chart: 1

Distance:
12 million light-years

Diameter:
35,000 light-years

Galaxy type: I0

Black hole mass:
30 million x Sun

A Burning Cigar

Only five times as far away as the Andromeda Galaxy—in astronomical terms, just around the corner—there is a high-energy starburst galaxy, in which at least ten times more stars are being formed than in the Milky Way. The enormous star-forming activity in this galaxy, M82 (also known as the Cigar Galaxy because of its shape), is caused by gravitational disturbances that occurred a few hundred million years ago, when it passed close to the neighboring spiral galaxy M81.

From Earth, we see M82 almost exactly edge-on. Because of that—and also because of the chaotic clouds of gas and dust in the galaxy—we cannot clearly see its true structure. For a long time, the Cigar Galaxy was classified as "irregular," but we now know that it is a spiral galaxy with an extremely bright, active nucleus. The nucleus alone radiates five times as much energy as the entire Milky Way.

At the center of the galaxy is a supermassive black hole around 30 million times more massive than the Sun. The activity of the black hole and the high-energy radiation of hundreds of young star clusters blow interstellar gas out of the galaxy in two directions, perpendicular to the galaxy's disk. That expelled gas has been imaged using radio and X-ray telescopes.

As so many new, massive stars are born in M82, it should come as no surprise that many supernova explosions occur here: some ten each century. The most recent of these was discovered in January 2014. It was a type Ia explosion, caused by the detonation of a white dwarf star.

▼ Wisps of radio waves in the central region of M82 are produced by charged particles interacting with the galaxy's magnetic field.

▲ Infrared, optical, and X-ray observations of the Cigar Galaxy have been combined to produce this colorful composite picture.

▼ The lenticular galaxy NGC 524 still exhibits a faint remnant spiral structure, although it does not have gas-rich spiral arms.

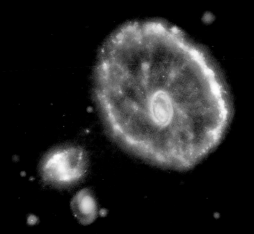

▲ Like lenticular galaxies, the Cartwheel Galaxy lacks spiral arms. In this case, the ringlike shape is the result of a collision with a companion.

Out of the Box

Wherever humankind tries to put nature "into boxes," intermediate forms crop up that completely mess up all our neat classification systems. *Dwarf planets* set on the fence between asteroids and "real" planets, *brown dwarf stars* are a kind of halfway house between stars and giant gas planets, and mysterious lenticular galaxies combine the properties of both spiral and elliptical galaxies.

Unlike elliptical galaxies, lenticular galaxies have a flattened disk in which most of their stars are concentrated, but they have no spiral arms and contain almost no interstellar gas. Consequently, very few new stars are formed in them, and the stars that are already there are mostly quite old, as is the case with elliptical galaxies.

Astronomers do not know exactly how lenticular galaxies originate. Roughly speaking, there are two hypotheses. According to the first, they are "extinguished" spiral galaxies; the gas in the galaxy is exhausted, so no new stars are born. Over time, the galaxy then loses it typical spiral structure. Some lenticular galaxies do, however, still have a central bar.

The second hypothesis says that lenticular galaxies are the product of two smaller galaxies colliding and merging. In the aftermath of the collision, there is a "baby boom" of new stars, but then the gas stocks really are exhausted. The Cartwheel Galaxy could be a good example to support this explanation. Lenticular galaxies do often contain some interstellar dust that can sometimes be seen in a rudimentary spiral pattern and sometimes, if we see the galaxy edge-on, as a dark band.

▼ NGC 4710 is a lenticular galaxy seen edge-on.

Galaxies

▼ The giant elliptical galaxy M87 lacks internal structure, apart from its conspicuous jet, which had already been observed in 1918.

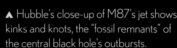

▲ Hubble's close-up of M87's jet shows kinks and knots, the "fossil remnants" of the central black hole's outbursts.

Galactic Champion

French astronomer Charles Messier had no idea that he was looking at a cosmic monster in 1781, when he discovered a nebulous spot in the constellation of Virgo. He recorded it as number 87 in his catalog of nebulous objects. We now know that it is a colossal elliptical galaxy, at a distance of about 54 million light-years from Earth.

M87 is one great collection of superlatives. It is about one million light-years across and weighs 200 times as much as the Milky Way. No less than 12,000 globular clusters swarm around the giant galaxy (the Milky Way has only a few hundred), and in the center of the colossus is a supermassive black hole 6.5 *billion* times more massive than the Sun.

American astronomer Heber Curtis discovered in 1918 that M87 has a striking "jet," an elongated streak of light pointing in one direction away from the center. In the mid-twentieth century, it was discovered that the light from the jet is polarized and that M87 produces large quantities of radio waves, suggesting that the jet is a protracted, magnetized stream of high-energy electrons ejected into space by the central black hole.

Besides visible light and radio waves, M87 also emits a lot of X-rays and gamma rays. The giant galaxy is located at the center of the Virgo Cluster, a gigantic agglomeration of galaxies. Gas from the cluster flows into the galaxy, most of which will in time be swallowed up by the central black hole.

◄ Radio telescopes have mapped giant lobes of high-energy electrons at huge distances from M87.

▲ High-resolution radio observations reveal short-term changes in the structure of the jet.

The Black Hole Inside

Astronomers discovered that there is a massive black hole at the center of the Milky Way only some 40 years ago. We now know that practically every galaxy in the Universe has a supermassive black hole in its nucleus. Sometimes, the gravitational effect of these dark monsters is visible from the velocities of stars close to the nucleus of the galaxy. In other cases, the presence of a supermassive black hole can be deduced from the activity of the nucleus in the form of radio waves and X-rays.

The black hole in the nucleus of the Milky Way is four million times more massive than the Sun. That seems a lot, but it is nothing compared with the supermassive black holes in some other galaxies. M87 has a black hole of 6.5 billion solar masses; the elliptical galaxies NGC 3842 and NGC 4889 (both around 330 million light-years away) contain black holes of about ten and twenty billion solar masses.

No one knows exactly how all these supermassive black holes originated. They may have been caused by the merger of countless "stellar" black holes produced by powerful supernova explosions. In any case, the growth of supermassive black holes seems to keep pace with that of the galaxies in which they are located: their mass is always around 0.1% of the mass of the galaxy (or of its central bulge).

Two galaxies have two supermassive black holes in their nucleus. These double black holes were probably created by the collision and merger of two smaller galaxies.

▼ Most supermassive black holes are surrounded by thick donuts of dark dust.

▲ Peculiar galaxies like Arp 220 (background photo) harbor a central supermassive black hole that spews jets of matter and radiation into space (artistic overlay).

◀ A black hole is surrounded by a swirling accretion disk of inwardly-falling matter in this artistic impression.

◀ Galaxy NGC 3842 (lower left) has a supermassive black hole weighing in at ten billion times the mass of the Sun.

Name: Centaurus A
NGC 5128

Constellation:
Centaurus

Sky position:
R.A. 13ʰ 25ᵐ 28ˢ
Dec. -43° 01.1′

Star chart: 11

Distance:
11 million light-years

Diameter:
100,000 light-years

Galaxy type: SO pec

Black hole mass:
55 million x Sun

Nearby Wreck

Of all the active galaxies in the Universe, NGC 5128 is the closest, at a distance of slightly more than 11 million light-years. It is a large galaxy, and thanks to its relative proximity, it can be easily seen using a small telescope. This is possible, however, only from the tropics or from the southern hemisphere, as NGC 5128 is located in the southern constellation of Centaurus.

Through a normal telescope, the galaxy looks like an elliptical or lenticular galaxy seen edge-on. The bright nucleus is dissected by a dark, slightly warped band of dust. The striking shape of the band and the countless young star clusters it contains suggest that NGC 5128 is the result of the collision and merger of smaller galaxies many hundreds of millions of years ago.

A radio telescope produces a completely different image of the galaxy, which radio astronomers call Centaurus A. Two enormous jets of rapidly moving electrons, more than a million light-years long, are being ejected from the nucleus in opposite directions. Where the material in the jets slams into the rarefied gas in intergalactic space, great "lobes" of radio waves are created.

Many of these radio galaxies have been discovered in the Universe, but Centaurus A is the closest to Earth and can therefore be studied in great detail. As with other radio galaxies, the jets originate in the immediate vicinity of a supermassive black hole in the nucleus of the galaxy. The black hole in Centaurus A is around 55 million times more massive than the Sun.

▲ A close-up of the galaxy's core reveals turbulent clouds of dust and bright star-forming regions.

> This long-exposure photograph shows how the outer regions of the elliptical galaxy Centaurus A extend far beyond its dusty center.

> At infrared wavelengths, NASA's Spitzer Space Telescope sees the glow of warm dust all the way into the giant galaxy's heart.

▲ NASA's Chandra X-ray Observatory stared at Centaurus A for 7 days to create this detailed view of the galaxy's powerful jets.

▲ Submillimeter observations (orange) and X-ray measurements (blue) are combined with an optical photo of Centaurus A to reveal its jets and lobes.

PASSPORT

Name: 3C273

Constellation: Virgo

Sky position:
R.A. 12ʰ 29ᵐ 07ˢ
Dec. +02° 03.1'

Star chart: 5

Distance:
2.4 billion light-years

Diameter:
300,000 light-years

Galaxy type: Sy1

Black hole mass:
900 million x Sun

Quasar Questions

At the end of the 1950s, radio astronomy was still in its infancy. The real nature of many cosmic sources of radio waves was unknown. At that time, even their position in the sky could not be determined precisely.

When astronomers did succeed in determining the precise sky position of 3C273 (number 273 in the *Third Cambridge Catalogue of Radio Sources*) in 1962, it at first seemed to be a small, faint star. But spectroscopic measurements by Dutch-American astronomer Maarten Schmidt showed that it must be almost 2.5 billion light-years away, which meant that it was a quasi-stellar object (QSO)—an extremely distant galaxy that generates a massive amount of energy.

We now know that such "quasars" (a contraction of "quasi-star") are in reality the very active nuclei of distant galaxies. The supermassive black hole in the nucleus of the galaxy swallows up large quantities of material, causing enormous amounts of gas and radiation to be blown into space. Like M87 and Centaurus A, 3C273 has an impressive jet and produces a lot of X-rays.

Such active galaxies were much more numerous when the Universe was young than they are now. How they appear to us depends largely on our viewing perspective. If we see them more or less edge-on, the extremely bright nucleus is almost or completely concealed by a thick "donut" of gas and dust, and we see a radio galaxy or a quasar. If we look straight into the jet, we see a bright "blazar."

➤ Close to the quasar's luminous core, the Hubble Space Telescope captured this nice view of 3C273's jet.

➤ Detailed radio observations of 3C273 (white dot at lower right) reveal that the jet ends in a larger lobe of radio emission (colored area).

◀ This artistic impression shows quasar ULAS J1120+0641, which is powered by a supermassive black hole two billion times as massive as the Sun.

➤ This is how the quasar-like core of the active galaxy NGC 3783 might look from nearby.

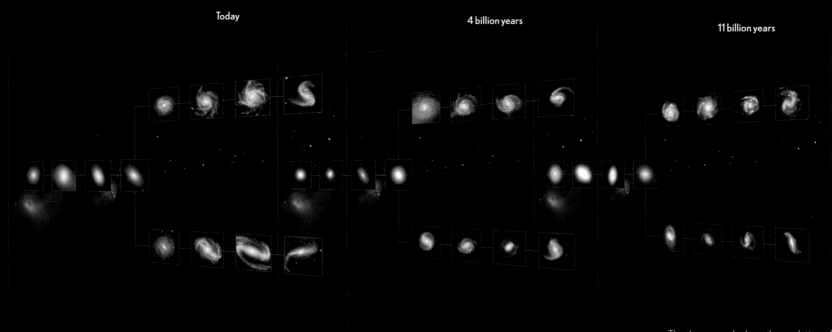

▲ The three panels show the evolution of elliptical, spiral, and barred spiral galaxies over time, from 11 billion years ago (right) to the present (left).

Galaxy Growth

With the Hubble Space Telescope, astronomers have discovered extremely faint objects at distances of many billions of light-years. That means they are also looking back billions of years back in time: the light from these distant galaxies must have been emitted when the Universe was still very young. Thanks to such "deep-field" observations, we have acquired a slightly better idea of how galaxies form and evolve, although the details are often still unclear.

The first galaxies must have been formed at most 100 million years after the Big Bang. They were often irregular in shape, much smaller and less massive than the Milky Way, and consisted primarily of extremely massive, rapidly evolving stars. In the course of billions of years, these small galaxies merged to form larger ones, which often developed a majestic spiral structure. Later, larger spiral galaxies collided to form giant elliptical galaxies.

And yet many questions remain unanswered. How is it possible that large, massive galaxies with giant black holes in their nuclei already existed so shortly after the Big Bang? Was the first energetic radiation in the early youth of the Universe produced primarily by stars or by the active nuclei of galaxies? And where are the hundreds of dwarf galaxies that, according to the best theories and computer simulations, should still be swarming around large spiral galaxies like the Milky Way?

It is hoped that many of these questions will be answered in the coming years by large, new telescopes like the Atacama Large Millimeter/Submillimeter Array in Chile and the James Webb Space Telescope, which will succeed Hubble.

▲ Even these small galaxies, observed by the Hubble Space Telescope in the very early Universe, appear to harbor supermassive black holes.

Windows on the Cosmos

The human eye is sensitive only to "visible" light, that is, electromagnetic radiation at wavelengths between 400 nanometers (violet) and 700 nanometers (red). It therefore took a long time to discover the existence of completely different types of radiation in the natural world. William Herschel came across infrared radiation by coincidence in 1800, ultraviolet radiation was discovered by Johann Ritter in 1801, and X-rays by Wilhelm Röntgen in 1895. All these others kinds of radiation are also emitted in the Universe. Studying only the visible wavelengths in the cosmos, therefore, produces a very limited picture. Many objects and phenomena cannot be observed in visible light. The emergence of radio astronomy halfway through the twentieth century, for example, made it possible for the first time to map large, cold clouds of neutral hydrogen gas in the Universe. Radio waves are also emitted by rapidly moving electrons in magnetic fields.

Millimeter and submillimeter radiation (between radio waves and infrared radiation) is produced by clouds of cold gas and dust. The Atacama Large Millimeter/submillimeter Array (ALMA) in Chile conducts measurements in these wavelengths to learn more about the birth of galaxies, stars, and planets. Complex molecules in the Universe can also best be studied at these wavelengths, as can the cosmic background radiation—the "echo" of the Big Bang.

Infrared radiation (heat radiation) is preferentially emitted by objects around room temperature. Astronomers use infrared telescopes on Earth and in space to study star-forming regions that are hidden from sight to optical telescopes by absorbing dust clouds. Infrared telescopes are also excellent for observing extremely distant galaxies.

High-energy, short-wave radiation from the Universe (ultraviolet light, X-rays, and gamma radiation) can be observed only from space: life on Earth is fortunately protected from these deadly rays by the atmosphere. Observations at ultraviolet wavelengths focus mainly on young, hot giant stars, white dwarfs, and "baby booms" of stars in other galaxies.

X-ray telescopes must be used to observe the hottest and most energetic phenomena in the Universe: rarefied, hot gas in interstellar space, matter swallowed up by black holes, rapidly spinning neutron stars, and supernova explosions. The last of these also generate gamma rays, the most energetic radiation in nature. Astronomers conduct observations at gamma wavelengths to study gamma-ray bursts, radioactive processes in the Universe, and the annihilation of matter and antimatter.

Besides electromagnetic radiation, the Universe also produces cosmic rays: high-energy electrically charged particles that come to Earth from space. Some of these subatomic particles have as much energy as a powerfully served tennis ball! It is not yet clear where they come from. The same applies to cosmic neutrinos, which can be detected only using special underground detectors.

One kind of cosmic signal has never yet been detected: gravitational waves. The existence of these minuscule vibrations of space–time is predicted by Einstein's theory of relativity but has been demonstrated only indirectly. Astronomers hope to use sensitive detectors like the Laser Interferometer Gravitational-wave Observatory (LIGO) in the United States and VIRGO in Europe to track these waves. In addition, work is under way to build a large space observatory to detect gravitational waves.

▲ Our Milky Way looks completely different at various wavelengths. The third panel from below shows the optical view.

▼ Radio telescopes reveal giant jets of high-energy particles on both sides of the galaxy Hercules A, also known as 3C348.

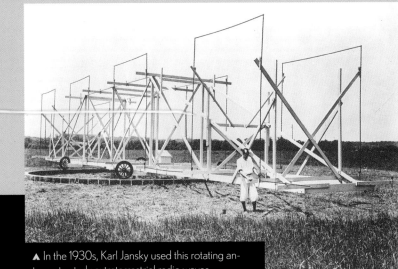

▲ In the 1930s, Karl Jansky used this rotating antenna to study extraterrestrial radio waves.

◀ Around 3,000 infrared Spitzer Space Telescope photos were used to create this mosaic of the Andromeda Galaxy, showing dust lanes in the spiral arms.

▼ In the star-forming region RCW120, the pressure of hot gas (orange) causes surrounding cooler material to clump, as shown here at submillimeter wavelengths.

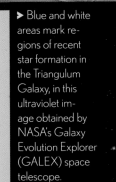

▼ The VIRGO gravitational wave observatory near Pisa, Italy, is on the hunt for minute ripples in space-time.

▲ Millimeter-wave observations of the Boomerang Nebula, a very young planetary nebula, indicate extremely low temperatures in the expanding gas.

▶ Blue and white areas mark regions of recent star formation in the Triangulum Galaxy, in this ultraviolet image obtained by NASA's Galaxy Evolution Explorer (GALEX) space telescope.

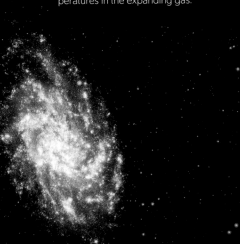

Windows on the Cosmos

CLUSTERS

George Abell must have regularly had tired, sore eyes. As part of his doctoral research in 1957, the American astronomer spent many long hours bent over the photographic plates of the Palomar Sky Survey, which had recorded millions of stars and faint galaxies. Back then, there were no smart computers, so Abell had to scour the images with the naked eye, looking for clusters of galaxies: groups of tens or thousands of galaxies that were close together in the sky and about the same distance from the Earth. In 1958, Abell published his "Northern Survey," a catalog of 2,712 clusters. In the 1970s, the catalog was supplemented by the "Southern Survey," which listed another 1,361 clusters.

Abell's groundbreaking work showed that galaxies are not regularly distributed throughout the Universe. They are part of small groups that in turn form larger clusters, and these clusters are often themselves grouped into elongated *superclusters*. The vast spaces between these superclusters are unimaginably empty *supervoids* containing almost no galaxies at all.

Thanks to large-scale observation campaigns using powerful telescopes and sensitive spectrographs, astronomers have mapped the "soapsuds" structure of the Universe. Measurements using gravitational lenses have even enabled them to determine the distribution of mysterious dark matter.

Studying clusters and superclusters is essential to acquiring a greater insight into how the Universe evolved. Its current large-scale structure originated in minute density variations in the hot gas that filled the expanding Universe shortly after the Big Bang.

◄ In this Hubble Space Telescope picture, images of background galaxies are magnified and distorted by the gravity of Abell 2744, a remote cluster of galaxies.

Name: Virgo Cluster

Constellation: Virgo

Sky position:
R.A. 12^h 27^m
Dec. +12° 43'

Star chart: 5

Distance:
55 million light-years

Diameter:
15 million light-years

Mass: 1.2 x 10^{15} x Sun

Number of galaxies:
1,500

▶ The Virgo Cluster is a huge collection of at least 1,500 galaxies, including the bright ellipticals M49 and M87.

▶ Black spots are used to blot out the light of foreground stars in this long-exposure photograph, which reveals the true extent of the faint galaxy halos.

▲ Close to the giant elliptical galaxy M60 is a smaller spiral known as Arp 116. Both galaxies belong to the Virgo Cluster.

The Virgo Concourse

The largest "nearby" cluster of galaxies is in the constellation of Virgo. In 1781, Charles Messier noted that there were a lot of "nebulae" in this part of the sky, including two bright ones, which he designated Messier 49 (M49) and M87. The term *Virgo Cluster* was first used in 1931 by American astronomer Edwin Hubble. Today, we know that the cluster contains at least 1,500 individual galaxies and is located some 50 to 60 million light-years from Earth.

Interestingly, as in most other clusters of galaxies, the space between the individual galaxies in the Virgo Cluster is not really empty. The cluster, which is about 15 million light-years across, is filled with a rarefied gas—mostly hydrogen and helium—with a temperature of around 30 million degrees. That extremely hot gas can only be imaged using X-ray telescopes. In addition, large numbers of solitary stars and planetary nebulae have been found in the space between the galaxies; it is estimated that one in ten stars in the Virgo Cluster do not belong to a galaxy!

Precision measurements of the distances and motions of the many hundreds of galaxies in the cluster have provided information on the spatial structure of the Virgo Cluster. The spiral galaxies are spread out in an elongated "cloud" that more or less points in the direction of the Milky Way; the large, massive elliptical galaxies populate a more spherical region near the center of the cluster. The Virgo Cluster is also part of the Virgo Supercluster, to which the Local Group—the Milky Way's "backyard"—also belongs.

▲ Both the Local Group of galaxies, home to our own Milky Way (top), and the Virgo Cluster belong to the elongated Virgo Supercluster.

PASSPORT

Name: Coma Cluster
Abell 1656

Constellation:
Coma Berenices

Sky position:
R.A. 13^h 00^m
Dec. +27° 59'

Star chart: 5

Distance:
320 million light-years

Diameter:
20 million light-years

Mass: 7 x 10^{14} x Sun

Number of galaxies:
1,000

A Crowd in Coma

▼ The Hubble Space Telescope peered into the heart of the Coma Cluster at a distance of 320 million light-years.

Six times farther away than the Virgo Cluster is the Coma Cluster, also named after the constellation in which it is located (Coma Berenices). American astronomer George Abell, who compiled a catalog of "rich" clusters at the end of the 1950s, recorded the Coma Cluster as A1656. Like the Virgo Cluster, it contains more than 1,000 individual, mainly large elliptical galaxies.

Close to the center of the Coma Cluster are two giant elliptical galaxies, New General Catalogue (NGC) 4874 and NGC 4889. Both are surrounded by an enormous number of globular clusters (the same is true for the giant elliptical galaxy M87 in the Virgo Cluster). These clusters contain mostly extremely old stars, which were probably formed before the galaxies themselves, but their exact role in the evolution of the clusters is unclear.

supercluster. The Coma Cluster and the neighboring Leo Cluster, for example, together form the Coma Supercluster; and the Virgo Cluster is the central component of an elongated supercluster to which—right out in the outermost regions—the Local Group also belongs. The Virgo Supercluster contains an estimated 5,000 galaxies.

Astronomers have discovered hundreds more superclusters at much greater distances. In most cases, they are elongated chains of smaller clusters. Some 130 superclusters have already been found within a radius of 1.3 billion light-years. As this is only about 10% of the radius of the observable Universe, the Universe as a whole must contain around 100,000 superclusters.

Large superclusters "closer" to home include the Perseus-Pisces Supercluster and the Hydra-Centaurus Supercluster. The latter contains the "Great Attractor," a colossal concentration of material toward which the Virgo Supercluster, including our own Milky Way, is "falling" at a velocity of 600 kilometers per second.

The largest structures of all in the Universe are gigantic, elongated "walls" of hundreds of thousands of galaxies. The Great Wall, for example, discovered in 1989 at a distance of about 200 million light-years from the Earth, is more than half a billion light-years long. The Sloan Great Wall is even bigger, at 1.4 billion light-years, and the Hercules-Corona Borealis Great Wall, discovered in 2013 on the basis of the spatial distribution of distant gamma-ray bursts, is longer than ten billion light-years.

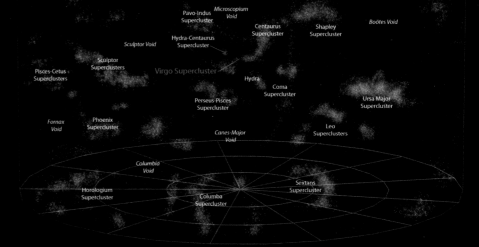

▲ Our Milky Way galaxy is much too small to be seen in this three-dimensional view of the distribution of superclusters in the Universe.

▼ The Sloan Telescope at Apache Point Observatory in New Mexico was used to discover the Sloan Great Wall, the largest known structure in the Universe.

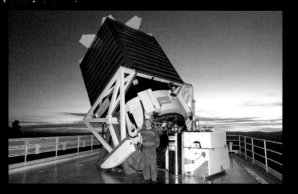

▲ Most galaxies in this image belong to the supercluster Abell 901/902, which spans 16 million light-years, at a distance of more than two billion light-years.

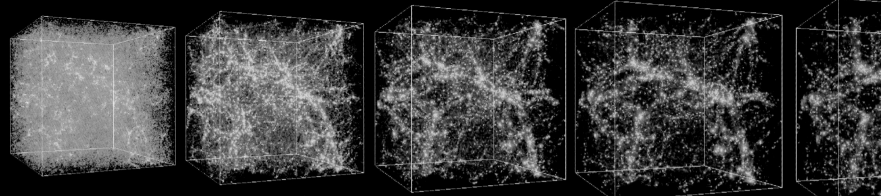

▲ Stills from a computer simulation show how gas in an expanding Universe clumps together to form the large-scale structure we observe today.

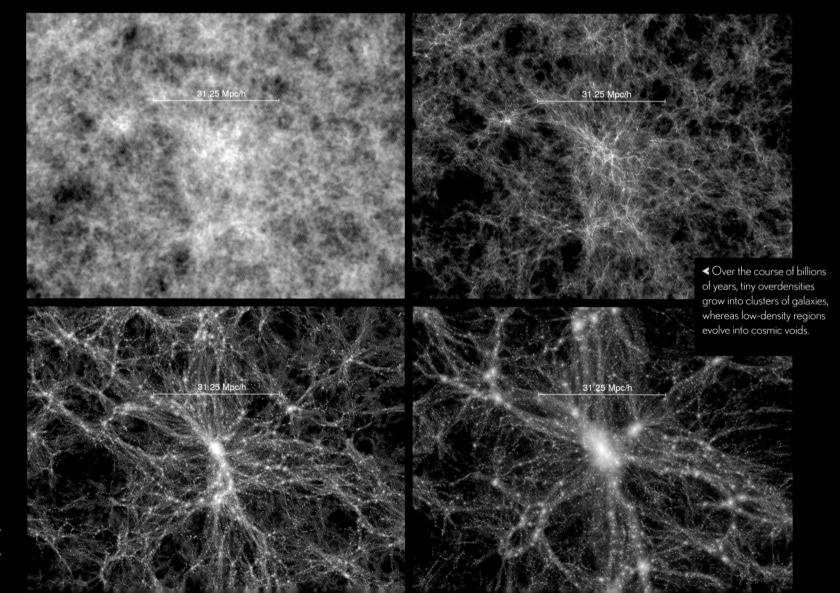

◄ Over the course of billions of years, tiny overdensities grow into clusters of galaxies, whereas low-density regions evolve into cosmic voids.

Deep Space

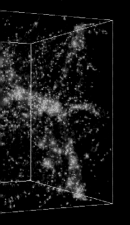

Soapsuds of the Universe

In the middle of the last century, George Abell discovered thousands of clusters by studying photographs of the night sky and looking for concentrations of galaxies, but to be certain that all these galaxies are genuinely clustered together in space, it is important to know how far away they are. In recent decades, thanks to extensive survey programs, the distances to hundreds of thousands of galaxies have been determined, and cosmologists now have a clear picture of the three-dimensional, large-scale structure of the Universe.

This spatial distribution most closely resembles soapsuds. The Universe is largely empty space, in the form of colossal *supervoids*, the "soap bubbles," which contain almost no galaxies. These more or less spherical voids are surrounded by thin "membranes" of galaxies. Where two of these membranes touch, elongated superclusters like the Great Wall are visible; and where these filaments cross lie the "rich" clusters mapped by George Abell.

This observed large-scale structure corresponds closely to computer-simulation predictions of the evolution of the Universe. Mysterious dark matter plays an important role in this process. It was primarily this dark matter that began to clump together in the expanding Universe under the influence of its own gravity. Only when these density concentrations were high enough was sufficient "normal" material also drawn together to form galaxies. The most fascinating idea is perhaps that the large-scale structure of the Universe originated in incredibly small quantum fluctuations, shortly after the Big Bang.

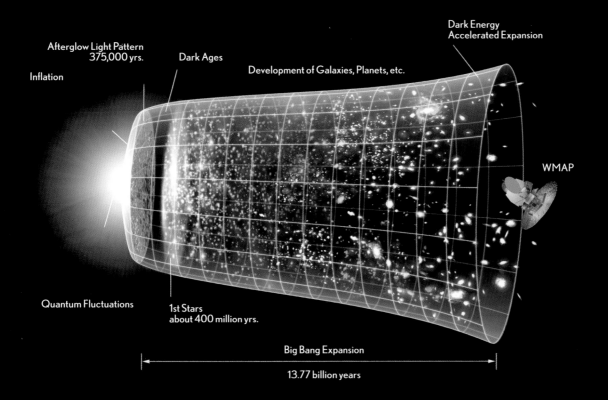

Afterglow Light Pattern 375,000 yrs.

Inflation

Dark Ages

Development of Galaxies, Planets, etc.

Dark Energy Accelerated Expansion

WMAP

Quantum Fluctuations

1st Stars about 400 million yrs.

Big Bang Expansion

13.77 billion years

◀ NASA's Wilkinson Microwave Anisotropy Probe (WMAP) satellite studied cosmic evolution, from the Big Bang, some 13.8 billion years ago, to the present-day Universe.

Distorted Views

Clusters of galaxies are fascinating enough in themselves, but astronomers also use them in a completely different way: as natural telescopes. In short, the gravity of a cluster distorts the light of more distant galaxies. We can use these "gravitational lenses" to study galaxies that would otherwise be too far away to be visible.

Albert Einstein predicted as early as 1916 that light would bend in a strong gravitational field and, 3 years later, English astronomer Sir Arthur Eddington showed this prediction to be true. The first gravitational lens was not discovered until 1979, however, when the image of a distant quasar was intensified and split into two by the gravity of a closer galaxy. Since then, many hundreds of gravitational lenses have been identified.

Looking at distant background objects through a cluster of galaxies is like looking at the world through pebbled glass. The image of such a faint galaxy is split into several components, all of which are stretched out to become long arcs of light. The brightness of the images is also increased by the gravitational lens so that (with large telescopes) they can be studied in detail. The new Hubble Space Telescope Frontier Fields program, for example, has been specially designed to carry out these kinds of observations. Study of gravitational lenses and light arcs provides a lot of information not only on the most distant galaxies in the Universe but also on the distribution of dark matter in the cluster, which can be mapped in detail by examining closely how the lens works.

◄ At a distance of 2.1 billion light-years, galaxy cluster Abell 2218 acts as a gravitational lens, magnifying and distorting the images of background galaxies.

◄ If the alignment is near perfect, the image of a background object can be stretched into an Einstein ring by the gravity of a foreground galaxy.

► The giant arc in this Hubble photograph is a galaxy at a distance of 10 billion light-years, strongly lensed by a cluster which is twice as close.

► In a gravitational lens, multiple images of a distant galaxy are produced by the light-bending gravity of a foreground object.

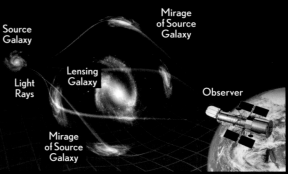

Source Galaxy

Mirage of Source Galaxy

Lensing Galaxy

Light Rays

Observer

Mirage of Source Galaxy

Biting the Bullet

Besides "strong lensing," clusters can also have a "weak lensing" effect. The images of distant background galaxies are then only slightly distorted by the complex gravitational field through which the light passes during its multibillion-year journey to the Earth. Through statistical research into the shape and orientation of thousands of faint background galaxies, astronomers can determine the distribution of dark matter in space.

This method has been used to study the Bullet Cluster, which is more than 3.5 billion light-years from Earth. It was already known that it had been created by the collision of two clusters some 150 million years ago, when the smaller cluster passed practically through the center of the larger one. In most cases, the individual galaxies simply flew past each other, but this collision caused the hot intracluster gas between the galaxies to accumulate halfway between the two clusters. This can be seen with X-ray telescopes.

By conducting measurements of the weak lensing effect of the Bullet Cluster on distant background galaxies, it was possible to map the distribution of dark matter in the double cluster. This mapping showed that the dark matter is closely linked to the visible galaxies, just as most theories predicted.

This method has since been used to determine the spatial distribution of dark matter in other parts of the Universe. In some cases, astronomers have even been able to produce accurate three-dimensional maps. They hope to use this kind of research to help them learn more about dark matter.

▼ Hot X-ray emitting gas (pink) is distributed very differently from dark matter (blue) in this composite image of the Bullet Cluster.

◄ Abell 520 is a merging galaxy cluster similar to the Bullet Cluster. However, in this case, dark matter (blue) appears to have piled up in the center.

◄ The distribution of dark matter in galaxy cluster Abell 1689, as derived from measurements of weak lensing.

▶ Thousands of distant galaxies are visible in this Hubble image of a minute fraction of the sky. Our observable Universe contains some hundred billion galaxies.

The
UNIVERSE

Cosmology, the study of the birth, evolution, and structure of the Universe, is by definition the most all-encompassing of all the sciences. Of course, it appeals enormously to the imagination. Cosmology addresses questions that sat, for many centuries, firmly in the domain of religion and philosophy, such as, how did everything begin, and what is the place of humankind in this great Universe of ours?

It was not until the start of the twentieth century that fables, myths, and speculation about the origin and evolution of the Universe gradually made way for scientific insights based on objective observations. In recent years, these observations have become so accurate that some astronomers even speak of "precision cosmology."

Yet this term seems a little premature. No matter how much we now know about the large-scale structure of the Universe, the earliest evolution of galaxies and the cosmic background radiation—the "echo" of the Big Bang—we are still completely in the dark about the true nature of 96% of the content of the Universe, and no one knows whether there are countless other universes beyond (or parallel to) our own.

No one even suggests that we may one day find all the answers; perhaps the human brain is simply not equipped to completely understand the cosmos. But one thing is clear: humankind is an inseparable part of the Universe, and we cannot see our own existence in isolation from the billions of years the Universe has taken to evolve.

Einstein's 4D-Universe

A few hundred years ago, the Universe was still easy to understand. Isaac Newton, according to some the greatest scientist of the seventeenth century, saw space and time as an absolute, immutable backdrop against which all natural events and phenomena took place. *Space* was an endless void with three coordinates (front/rear, left/right, and up/down), which can best be compared to a three-dimensional piece of squared paper; and *time* was a kind of absolute clock, a time line along which the present moved second by second from the past to the future. In Newton's cosmos, both space and time existed entirely independently, even if the Universe was completely empty and nothing at all ever happened.

In the early twentieth century, that intuitive perception had to make way for Albert Einstein's notion that space and time are relative. In Einstein's view, space and time are irrevocably linked to each other, and neither is absolute. Distances, periods of time, even simultaneity are all relative concepts that mean something completely different to one observer than to another. Furthermore, they are influenced not only by the motion of the observer but also by the presence of matter; it is matter that determines whether and to what extent four-dimensional space–time is curved and distorted. What we experience as gravity is in fact synonymous with Einstein's notion of space–time being distorted. This has not made it easier to understand space and time, but it has made our conceptions much more versatile. Without Einstein's "malleable" four-dimensional space–time, we would never have come across such phenomena as cosmic expansion, black holes, gravitational waves, or hypothetical wormholes.

▲ Infinite three-dimensional empty space and constantly ticking never-ending time: that was the seventeenth-century view of Sir Isaac Newton.

▲ In Einstein's view, the precise shape of space-time is defined by the gravity of both visible and dark matter. As a result, light rays follow "wiggly" paths.

◄ Albert Einstein's general theory of relativity is our current best description of space-time. When he formulated the theory, Einstein was 36 years old, much younger than in this photograph.

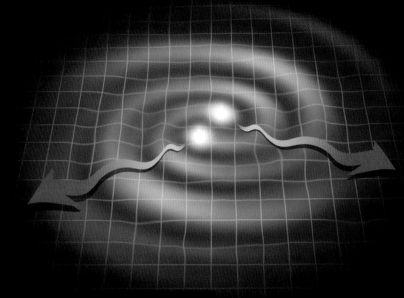

Space-Time Ripples

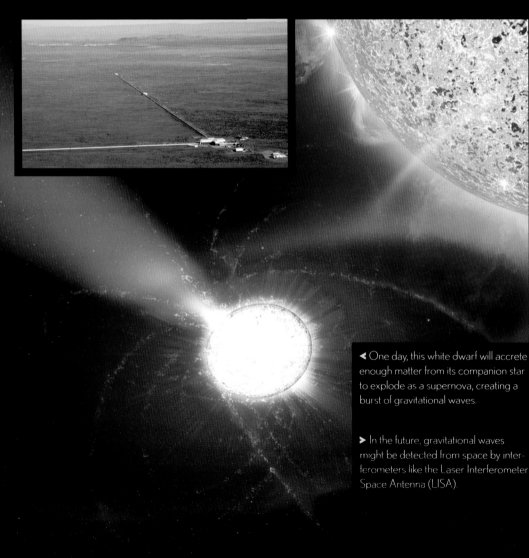

▼ The Laser Interferometer Gravitational-Wave Observatory (LIGO) in the United States consists of two identical setups, in Louisiana and in Washington State.

▲ When two neutron stars orbit each other, they lose energy in the form of gravitational waves.

◄ One day, this white dwarf will accrete enough matter from its companion star to explode as a supernova, creating a burst of gravitational waves.

► In the future, gravitational waves might be detected from space by interferometers like the Laser Interferometer Space Antenna (LISA).

Einstein's general theory of relativity predicts the existence not only of curved space and black holes, but also of *gravitational waves*—minuscule ripples in the structure of space–time. Just as Jell-O or church bells can be made to vibrate because they are, to different degrees, malleable, space–time can also tremble and resound, even though this requires enormous quantities of energy.

Every mass that changes its velocity or direction creates a gravity wave. In by far the most cases, this phenomenon passes by completely unnoticed. Only in extreme circumstances, for example, when two compact neutron stars circle each other or in the case of a supernova explosion, are gravitational waves generated that can just about be detected on Earth with our current technology.

Convincing indirect evidence for the existence of gravitational waves emerged in 1974, when Joseph Taylor and Russell Hulse discovered that the binary neutron star BR 1913+16 is losing orbital energy exactly as Einstein predicted: the "lost" energy is being radiated in the form of gravitational waves. However, despite unimaginably sensitive detectors being built in the United States, Europe, Japan, and Australia to record these minuscule ripples in space–time, gravitational waves have so far never been directly observed. Who knows, perhaps with a new generation of detectors, or with instruments in space, that may become a reality in the near future.

In March 2014, astronomers announced the discovery of gravitational waves generated in the very first moments of the life of the Universe. If confirmed, this would shed light on the phenomena that occurred immediately after the Big Bang.

Space Generation

Few topics in astronomy cause as much confusion as the expansion of the Universe. Much of that confusion could be avoided if, rather than the expanding Universe, we were to talk about expanding space.

All too often, cosmic expansion is described in terms of the galaxies in the Universe moving apart. That is correct to the extent that the distances between galaxies are indeed increasing; but this does not mean that they are physically moving through space. Rather, it is space *itself* that is expanding, so that more and more space is being generated. The galaxies are like spots on the surface of a party balloon: when the balloon is inflated, the spots move farther apart without actually moving across the surface of the balloon.

The idea that space has to expand *into* something is also incorrect. An infinite line can be stretched out, so that marker points along it move farther apart from each other. In the same way, an infinite three-dimensional space can also expand; it does not require an environment to expand into.

Lastly, it is a misconception to think we can observe the expansion of the Universe in our immediate surroundings. Because the structure of space–time is so closely related to the presence of matter, the expansion of space plays no role where gravity dominates, such as in our own solar system or in the Milky Way.

◀ Edwin Hubble studied spectra of distant galaxies and concluded that they appear to recede faster when they are farther away.

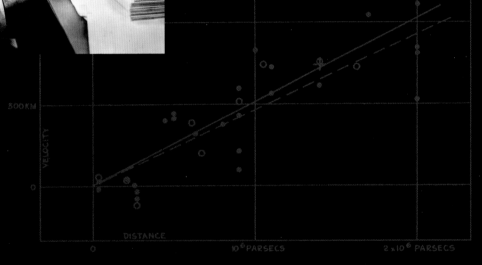

▼ This diagram, published by Edwin Hubble in 1929, was the first to show the relationship between the distance and the "recession velocity" of galaxies.

➤ Not only is the Universe expanding, but mysterious dark energy causes it to expand at an ever-increasing rate.

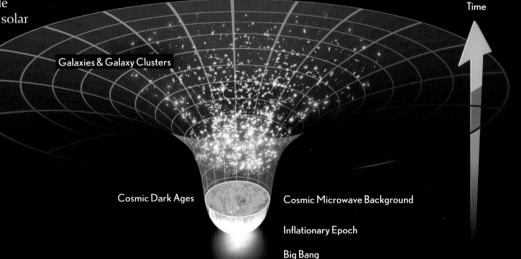

Dark Energy Accelerated Expansion

Time

Galaxies & Galaxy Clusters

Cosmic Dark Ages

Cosmic Microwave Background

Inflationary Epoch

Big Bang

Looking Back in Time

Astronomers are time travelers. Their telescopes do not show the Universe as it *is*, but how it *was*. Every time we look at the stars in the sky, we are looking into the past; and the farther we look into space, the farther we look back in time.

It is a fascinating thought that this enables us to see things that are no longer there: stars that exploded long ago as supernovae, galaxies that have since been ripped apart by tidal forces, the early youth of the Universe. This is all possible because light has a finite speed (300,000 kilometers per second) and therefore takes time to reach us.

The light from the Sun takes a little longer than 8 minutes to reach the Earth. That means we see the Sun as it was 8 minutes ago. But when we look up at the stars in the night sky, we are looking back in time much further, tens or hundreds of years, and astronomers use large telescopes to study galaxies that are so far away that their light has taken billions of years to get here.

Because we are looking back so far in time, the concept of "distance" loses some of its meaning in an expanding Universe. When astronomers say that a galaxy is 10 billion light-years away, they mean that its light has traveled through expanding space for 10 billion years before reaching us. When the light was emitted, the distance was much smaller and at the moment the light arrives at the Earth, the galaxy's distance is much larger.

▲ We do not see the Sun as it is right now, but as it was 8.3 minutes ago, the time it takes sunlight to reach Earth.

▼ The light that we receive today from NGC 1232 left the galaxy 60 million years ago, shortly after the extinction of the dinosaurs.

➤ At a distance of some 6,400 light-years, the Jewel Box Cluster is observed as it was in the year 4,400 BC.

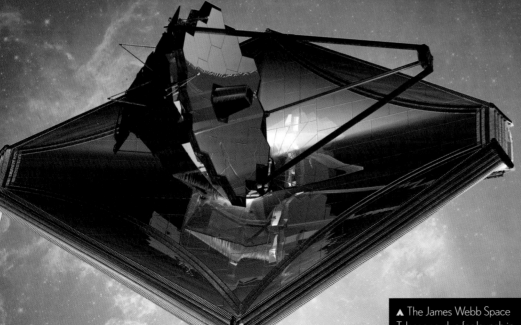

▲ Extremely remote galaxies (circled in this Hubble image) really appear redder than they are because of their huge redshifts.

Changing Colors

Thanks to the expansion of the Universe, astronomers are able to determine the distances of faraway galaxies. The light from a distant galaxy bears information about how long it has traveled through expanding space.

Every light wave has a specific wavelength. Red light, for example, has a longer wavelength than blue light, and infrared waves are even longer. Under normal circumstances, that wavelength does not change, and a light wave arrives at Earth with the same color it had when it was emitted; but in an expanding Universe, that is not the case.

The longer a light wave is en route, the more it is stretched by the expansion of the space through which it is moving. The light from a very remote galaxy takes longer to reach us than that from a nearby galaxy, and its light waves will therefore be stretched to a greater degree. When they eventually arrive at the Earth, the light waves will have a slightly redder color. This "redshift" in the light from galaxies can be measured using a spectroscope and is directly related to the time the light has travelled, and therefore to the distance to its source.

The light from the most distant galaxies in the Universe, which we therefore see as they were shortly after the Big Bang, has such a high redshift that hot, blue-white stars are almost impossible to see with optical telescopes like the Hubble Space Telescope and must be studied with an infrared telescope. That is why astronomers await the launch of the James Webb Space Telescope with such great anticipation. This successor to Hubble will primarily conduct observations at infrared wavelengths.

▲ The James Webb Space Telescope, due for launch in 2018, will study distant galaxies at infrared wavelengths.

The Observable Universe

The farther astronomers peer into boundless space, the further they look back in time. That is because the speed of light is not infinite. But that finite speed of light, together with the finite age of the Universe, has another consequence: there is a fundamental limit to what we can see. This "cosmological horizon" has nothing to do with the size of our telescopes.

Light needs time to reach us, but the available traveling time is not endless, for the simple reason that the Universe has existed "only" for some 13.8 billion years. The light from a galaxy at a distance of 20 billion light-years has therefore simply not had enough time to reach Earth. That means it is beyond our cosmological horizon.

The *cosmological horizon* is a little like the horizon a sailor sees from the crow's nest of a ship. Just as the ocean continues beyond the horizon, the Universe continues past the cosmological horizon. We are simply unable to see it.

How can galaxies in an expanding Universe with an age of 13.8 billion years be farther apart from each other than 13.8 billion light-years? Does that mean that the Universe is expanding faster than the speed of light? In fact, yes; but that does not contradict Einstein's theory of relativity: it is *space* itself that is expanding; there is no matter moving *through* space faster than light.

➤ No, this is not how you should imagine the Big Bang. After all, it was not an explosion *in* space, but an explosion *of* space.

▲ About a hundred billion galaxies are in the observable Universe, each containing tens or hundreds of billions of stars.

◄ This open cluster in the Eagle Nebula is well within our cosmological horizon. Beyond the horizon may be countless similar clusters.

In the Beginning

As with the expansion of the Universe, there are many misconceptions about the Big Bang. It's time to put them right, one at a time.

Misconception 1: The Big Bang was an explosion somewhere in empty space.

In fact, in a certain sense, the Big Bang occurred *everywhere* in the Universe; 13.8 billion years ago, density and temperature were inconceivably high at every point in the Universe. There is therefore no "central point" from which everything originated.

Misconception 2: At the moment when the Big Bang occurred, the whole Universe was concentrated in one point.

This is also not the case. Everything seems to suggest that the Universe extends infinitely into space, and if that is the case *now*, it was also the case 13.8 billion years ago. That infinite space was much more "compact" then, so the matter in the Universe had an extremely high density (and temperature).

Misconception 3: The Big Bang marks the point at which time began: the moment described as t = 0.

This is also incorrect. For a start, when astronomers speak of the Big Bang, they are not referring to the actual moment at which the Universe came into existence (we still know too little about that with any degree of certainty), but to the extreme circumstances in which the Universe found itself when it was a tiny fraction of a second old. Moreover, it is by no means completely certain that time started with the Big Bang; other theories state that time existed before our Universe existed.

▲ Belgian astronomer and Jesuit priest Georges Lemaître was the first to come up with the idea of the Big Bang.

▲ The fifth European Automated Transfer Vehicle (ATV-5), used as a freighter to the International Space Station (ISS), was called Georges Lemaître after the "Father of the Big Bang."

Echoes of the Big Bang

The whole Universe is still bathing in that primordial radiation, but it is now much dimmer and less energetic. As a result of space expanding (by a factor of approximately 1,000 since the radiation was emitted), the wavelength of the primordial glow has been stretched to around 1 millimeter and its temperature has fallen to only 2.73°C above absolute zero.

This cosmic background radiation was not discovered until 1965, and then more or less by chance. Since then the "afterglow" of the Big Bang has been studied in great detail using sensitive instruments on Earth and in space. The best map of the cosmic background radiation currently available to astronomers was produced by the European space telescope Planck in 2013.

The minuscule temperature variations in this "baby photo" of the cosmos—hardly more than a ten-thousandth of a degree—are caused by small variations in density in the newborn Universe. These later led to the formation of galaxies. It should also be possible to find traces of gravitational waves from the first fraction of a second in the background radiation. Some cosmologists even believe that we will someday find subtle patterns in the "fossil radiation" of the Universe that will point to the existence of other, "parallel" Universes.

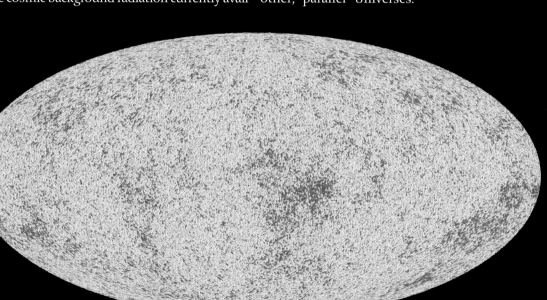

▲ The most detailed "baby photo" of the Universe so far was obtained by the European Planck spacecraft.

▶ In the early 1990s, NASA's Cosmic Background Explorer (COBE) satellite was the first to detect tiny temperature variations in the cosmic background radiation.

➤ The cosmic background radiation was discovered in 1965 by radio engineers Arno Penzias and Robert Wilson.

➤ In 2001, the Wilkinson Micro-wave Anisotropy Probe (WMAP) was launched to study the cosmic background radiation in more detail than COBE did.

> Ever since the Big Bang, the chemical complexity of the Universe has increased, leading to the formation of stars, planets, and life.

▼ By studying the very first galaxies in the Universe, the James Webb Space Telescope will shed light on our cosmic origins.

Deep Space

Cosmic Evolution

Thanks to precision measurements of the cosmic background radiation, of newborn galaxies many billions of light-years away, and of the large-scale structure of the Universe, with all its clusters and superclusters, astronomers have acquired a relatively good picture of the evolution of the Universe. Many details are still quite unclear, and the birth of the cosmos remains an unsolved mystery, but the main outlines of the story are solid as a rock. And it is perhaps the most wonderful story in the whole of science.

The newborn Universe was filled with a mixture of hydrogen and helium gas with an extremely high density and at an extremely high temperature. As space expanded, the gas became ever more rarefied and cooler. Small-density variations in the gas—probably "enlarged" quantum fluctuations generated immediately after the Universe's "birth"—grew under the influence of gravity into a spiderweb-like network of dark membranes, filaments, and clouds, separated by relatively empty voids. These were the seeds of the first galaxies.

Gravity also made itself felt on a small scale. Gas clouds collapsed under their own weight to form the first massive giant stars; and, for the first time in millions of years, the lights came on again in the Universe. Nuclear fusion reactions produced new elements, including carbon, oxygen, and nitrogen. Supernova explosions blew these heavier elements out into space. New generations of stars evolved from that stellar dust, many of them accompanied by planets. On at least one of those planets, organic molecules merged to form living cells. A few billion years later, *Homo sapiens* started building temples and telescopes.

▲ It took life more than three billion years to evolve from the first living cells to intelligent, self-conscious primates.

◄ Three-billion-year-old stromatolites, produced by single-celled microorganisms, are among the oldest fossil remnants of life on Earth.

Mystery Stuff

In the 1930s, astronomers discovered that there is more matter in the Universe than can be seen with telescopes. Measurements of the velocities of stars in the Milky Way and of individual galaxies in clusters show that there must be large quantities of dark matter. This mysterious stuff produces no radiation, but it does exert a gravitational effect on its surroundings. Later, convincing evidence for the existence of dark matter was also found in the rotational velocities of the outer regions of galaxies.

A small part of this dark matter consists of black holes; burned-out stars and intergalactic filaments of cold, dark gas. But it has now become clear that the lion's share of dark matter is not even made up of normal atoms and molecules. It is probably composed of unknown elementary particles. Except for their gravitational pull, they do not interact at all with "normal" particles. We have therefore not yet been able to detect them. Cosmologists and particle physicists are completely in the dark, literally, about the true nature of dark matter.

The existence of dark matter has so far been proved only indirectly, on the basis of its gravitational effects. Is it possible that our ideas about gravity are incorrect and that dark matter does not exist at all? Advocates of Modified Newtonian Dynamics (MOND) believe we *are* mistaken and have been spectacularly successful in explaining the rotational velocities of galaxies using an alternative hypothesis in which gravity falls off more slowly with distance than Newton prescribes. Many other observations, however, are harder to explain using the MOND hypothesis. For the time being at least, dark matter—mysterious as it is— seems to offer the best explanation.

▲ Pandora's Cluster (Abell 2744) contains hot intracluster gas (pink) and huge amounts of dark matter (color-coded blue).

▼ The motions of individual galaxies in clusters reveal that these huge agglomerates contain much more mass than meets the eye.

➤ The outer parts of galaxies like NGC 6946 rotate much faster than expected, hinting at the existence of large amounts of invisible matter.

▲ By mapping slight distortions in the shapes of background galaxies, astronomers are able to deduce the distribution of dark matter in a foreground cluster.

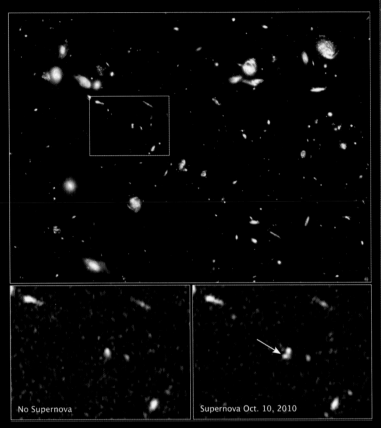

▲ The boxed region in the Hubble photograph at the top is shown in more detail at the bottom. In October 2010, a supernova flared up in a distant galaxy.

No Supernova

Supernova Oct. 10, 2010

▲ A white dwarf star (bottom) is about to explode as a type Ia supernova. Studies of type Ia's have revealed the expansion history of the Universe.

▲ Supernova SN UDS10Wil (boxed) is the farthest supernova ever detected. It exploded over 10 billion years ago.

Against Gravity

Anyone wishing to describe the content of the Universe must look not only at matter, but also at energy. According to Einstein's famous formula $E = mc^2$, matter and energy are two sides of the same coin. In 1998, astronomers discovered that the total matter–energy content of the cosmos is dominated by a mysterious sort of "antigravity," which made the expansion of space *accelerate* over time, rather than slow down.

The existence of this "dark energy" is clear not only from observations of distant supernovae, which, through redshift measurements, provide information about the history of the expansion of the Universe, but also from the properties of the cosmic background radiation. What kind of energy it is, whether it is related to the equally mysterious dark matter, and whether it is constant or also increases over times, no one knows.

The frustrating conclusion is that we actually have no idea of the composition of the cosmos; 74% of the total matter–energy content consists of mysterious dark energy, 22% is *nonbaryonic dark matter* (dark matter that does not consist of normal particles), the true nature of which we also do not know. Atoms and molecules account for only 4% of the total, and roughly three-quarters of that is too cold and dark to be observed. Stars, nebulae, and galaxies—the main "protagonists" in this book—amount to only 1% of a dark, incomprehensible Universe.

▲ The stars and galaxies that we see in our Universe constitute only 1 percent of the total matter-energy content.

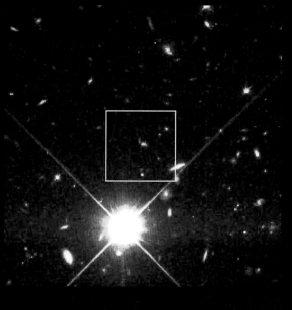

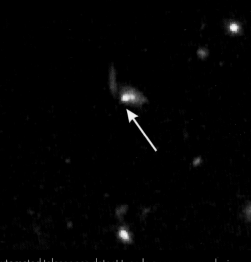

▲ Automated telescopes detect type Ia supernova explosions (arrowed) in extremely remote galaxies.

The Living Cosmos

Is Earth the only living planet in the Universe? We have been asking ourselves this question for centuries, but we still have no conclusive answer. However, given the unimaginable vastness of the Universe, in both space and time, it seems unlikely that life has evolved in only one place and at only one moment.

Until the early twentieth century, there was still serious speculation about the possibility of highly developed life-forms on our neighboring planet, Mars. We now know that life in our own solar system is pretty rare: Earth is teeming with life, but elsewhere there may be at most a few microorganisms, and even that is not certain. The discovery of planets orbiting other stars, however, has breathed new life into the debate on the possibilities of extraterrestrial life.

In star-forming regions and protoplanetary disks around newborn stars, astronomers have found not only molecules of water but also sugars and other organic compounds. These hydrocarbons are the fundamental building blocks of amino acids and of all life on Earth. What happened here on Earth more than three and a half billion years ago must also have been possible on many other planets in the Universe.

With the large telescopes of the future, like the James Webb Space Telescope in space, the Thirty Meter Telescope on Hawaii, and the European Extremely Large Telescope in Chile, we may be able to determine the composition of the atmosphere of an Earth-like exoplanet and may enable us to find traces of biological activity on the surface of the planet. Who knows, the discovery of extraterrestrial life may only be 20 or 30 years away.

▲ There must be billions of Earth-like planets in our Milky Way Galaxy alone. Some orbit binary or triple stars, leading to multiple sunsets.

▲ The organic building blocks of life have been discovered throughout the Universe.

◄ A couple billion years ago, Mars had oceans and a thicker atmosphere. Back then, it could have harbored life.

➤ Geysers at Jupiter's moon Europa indicate that there is a liquid ocean beneath the frozen crust.

Is There Anybody Out There?

For more than half a century, ever since the pioneering work by American radio astronomer Frank Drake, we have been searching for messages from aliens—artificial signals from extraterrestrial civilizations. It is not as crazy as it sounds: *Homo sapiens* also produce artificially generated radio waves that could be intercepted by the smart inhabitants of a distant exoplanet.

The current projects of the Search for Extraterrestrial Intelligence (SETI) are conducted with large radio telescopes like the one at Arecibo Observatory in Puerto Rico or with networks of smaller antennae, such as the Allen Telescope Array in California. They use extremely sensitive detectors that scan millions of radio frequencies and powerful supercomputers to process the data. Of course, special attention is devoted to the stars where planetary systems have been discovered. So far, however, the search has come up with nothing.

We also send messages out into space. Unmanned space probes that leave the solar system are carrying plaques and gramophone records with images and sounds from Earth, the cosmic equivalent of a message in a bottle. The National Aeronautics and Space Administration, or NASA, has broadcast The Beatles' "Across the Universe" in the direction of the Pole Star. Mathematically coded messages have been sent to nearby exoplanetary systems. We are screaming from the rooftops that we are here, but the cosmos remains eerily silent.

No one knows what that means. Perhaps aliens are silently watching us as a kind of scientific experiment. Perhaps their technology is so much more advanced than ours that we do not recognize their means of communication. Or perhaps they simply do not exist. Life is probably quite widespread in the Universe, but complex life-forms and intelligent civilizations might be extremely rare.

▲ The 305-meter Arecibo radio telescope in Puerto Rico is used to eavesdrop on alien radio communication.

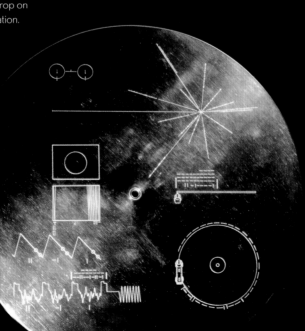

▲ In California, the privately funded Allen Telescope Array scans the skies at millions of frequencies, in search of extraterrestrial signals.

◀ Aliens may have a hard time deciphering the messages on the Voyager Golden Record, humankind's message in a bottle.

The Multiverse

Long ago, astronomers thought the Earth is unique. Now we know that seven other planets orbit our Sun. Once we thought the Sun is unique. Now we know that the Milky Way has hundreds of billions of suns. Only a century ago, most astronomers assumed that the Milky Way is unique. But we now know that the observable Universe contains at least 100 billion galaxies.

What about the Universe? Is it unique? Or are there other universes? If so, can we ever discover them?

It could hardly be called "hard science," but scientific journals are full of speculation about an enormous range of possible parallel universes. The popular string theory from particle physics actually seems to prescribe the existence of such a "multiverse." The different universes in this multiverse could each have their own characteristic properties, making our own, surprisingly "viable" Universe just a random ticket in a cosmic lottery.

Perhaps these different universes exist parallel to each other, in different dimensions. Or perhaps they succeed each other in time. It has even been suggested that each "mother universe" can give birth to a large number of "baby universes," so that there is a kind of cosmic genealogy. In an infinite universe, there may be an infinite number of different "domains" that are unimaginably far beyond our provincial cosmological horizon.

Whether we will ever find irrefutable evidence for the existence of parallel universes is doubtful; but the idea of a multiverse does give a completely new dimension to the concept of deep space.

◄ Like the Earth, the Sun and the Milky Way galaxy, our Universe may also not be unique. The cosmos we have learned to know could well be part of a never-ending multiverse.

Star Atlas

The following 14 pages present a star atlas showing the night sky as seen from Earth. The charts have been especially drawn for this book by Dutch astrocartographer Wil Tirion.

All the stars in the atlas are visible with the naked eye. A telescope is often required to see binary stars, nebulae, and galaxies. Most of the astronomical objects referred to in this book can be found in the star atlas.

The stars on Chart 1 are located in the area around the north celestial pole and are visible only from the northern hemisphere. For observers who live north of the Tropic of Cancer (approximately 23.5 degrees north), these stars never disappear below the horizon.

The stars on Charts 2 through 7 are also visible from the northern hemisphere but not throughout the year and mostly not for the whole night. Those on Charts 8 through 13 are visible from the southern hemisphere but also not at all times. The following table shows the periods of best visibility.

Lastly, the stars on Chart 14 are located around the south celestial pole and are visible only from the southern hemisphere. For observers who live south of the Tropic of Capricorn (approximately 23.5 degrees south), these stars never disappear below the horizon.

The coordinates of celestial objects in the sky are indicated by their "right ascension" and "declination." The right ascension in the night sky is comparable to latitude on the surface of the Earth. The celestial equator—the dividing line between the northern and southern hemispheres of the sky, which lies exactly above the terrestrial equator—is divided into 24 hours. The declination can be compared with the longitude on Earth's surface: the celestial equator has a declination of 0 degrees, and the north and south celestial poles have a declination of +90 degrees and –90 degrees, respectively.

Page 217 shows a complete list of the 88 official constellations recognized by the International Astronomical Union.

Period	Northern Hemisphere	Southern Hemisphere
October/November	Chart 2	Chart 8
December/January	Chart 3	Chart 9
February/March	Chart 4	Chart 10
April/May	Chart 5	Chart 11
June/July	Chart 6	Chart 12
August/September	Chart 7	Chart 13

LEGEND KEY

Stellar magnitudes:

- Brighter than –0.5
- –0.5 to 0.5
- 0.5 to 1.5
- 2.0
- 2.5
- 3.0
- 3.5
- 4.0
- 4.5
- 5.0
- 5.5
- 6.0

Double or multiple stars:

Variable stars:

Fainter than mag. 6.0 at minimum:

Milky Way:

Deepsky objects:
- Open star cluster
- Globular star cluster
- Planetary nebula
- Bright nebula
- Dark nebula
- Galaxy
- Quasar
- Radio or X-ray source
- Other interesting object

Constellation figures

Constellation boundaries

+10°   8ʰ   Grid lines with RA and dec

Equator

Celestial Equator

20°   *Ecliptic*

Ecliptic with longitude

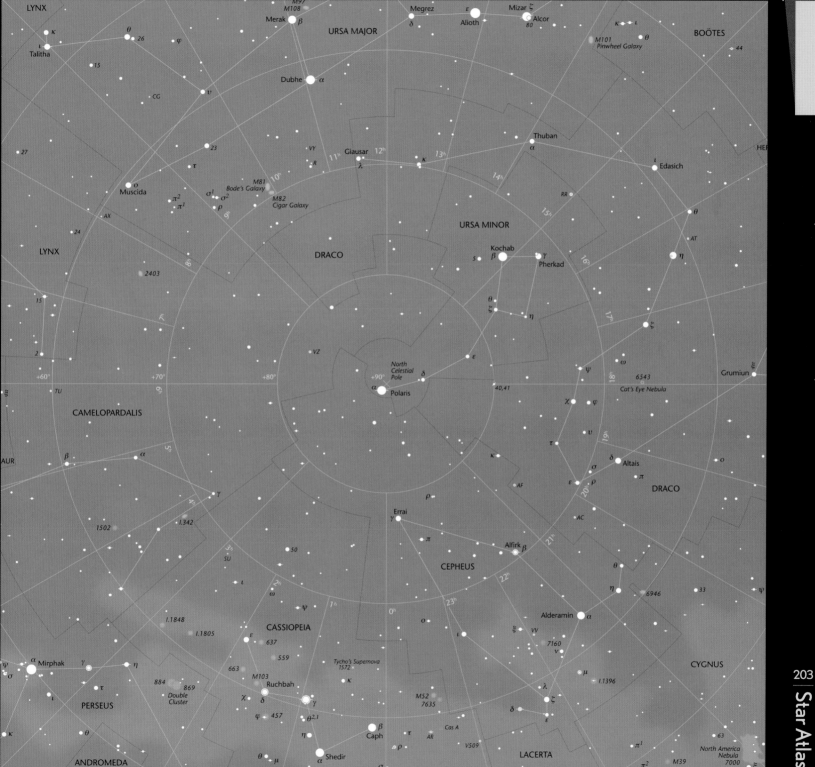

Chart 1

LYNX

M97
M108
Merak β   URSA MAJOR   Megrez ε   Mizar ξ
                                        δ   Alioth   Mizar ξ Alcor
                                                          80
κ   κ ι   BOÖTES
ι θ   26   M101   θ
Talitha   φ   Pinwheel Galaxy
                                                                                        44
15
                                        Dubhe α
                                                                Thuban
                                                                α
27   23   ι Edasich
           τ   VY   HER
              CG   R   11ʰ Giausar 12ʰ   κ 13ʰ
ο   λ   14ʰ
Muscida   M81   RR   θ
π² σ¹ Bode's Galaxy   AT
AX π¹ ρ σ² M82   10ʰ   URSA MINOR
24   ρ θ   Cigar Galaxy   η
LYNX   Kochab
                                                5 β γ
DRACO   Pherkad
2403   8ʰ   ξ
15   θ
                                                ξ η
2   7ʰ   ω
                                                                    ψ   6543
                              VZ   ε   χ φ Grumiun
+60°   TU   +70°   +80°   +90°   δ   North   40,41   Cat's Eye Nebula
                                    α Celestial Pole   υ
CAMELOPARDALIS   Polaris   τ
                                                                κ
β   α   6ʰ   δ Altais
                                                                σ
γ   ε 20ʰ ρ   π   DRACO
AUR   AC
1502   1.342   Errai   θ
1   γ   η
       SU   50   Alfirk β   6946   33   ψ
       ι   2ʰ   ο   Alderamin α
       ω   ξ VV
1.1848 ψ   1ʰ   0ʰ   23ʰ   22ʰ   CEPHEUS   7160   ν   CYGNUS
1.1805   μ
α Mirphak γ   η   CASSIOPEIA   ο   λ   1.1396
ψ σ   637   ι   ξ
τ   663 559   δ ε
884 869   M103   κ   M52
Double   Ruchbah χ δ γ θ²·¹   Tycho's Supernova 7635   Cas A
Cluster   φ 457   1572
PERSEUS   VS09   LACERTA
θ   η   β   τ   63 π¹
κ   θ μ   Caph   ρ   North America M39
ANDROMEDA   α Shedir   Nebula π²
                                                                          7000

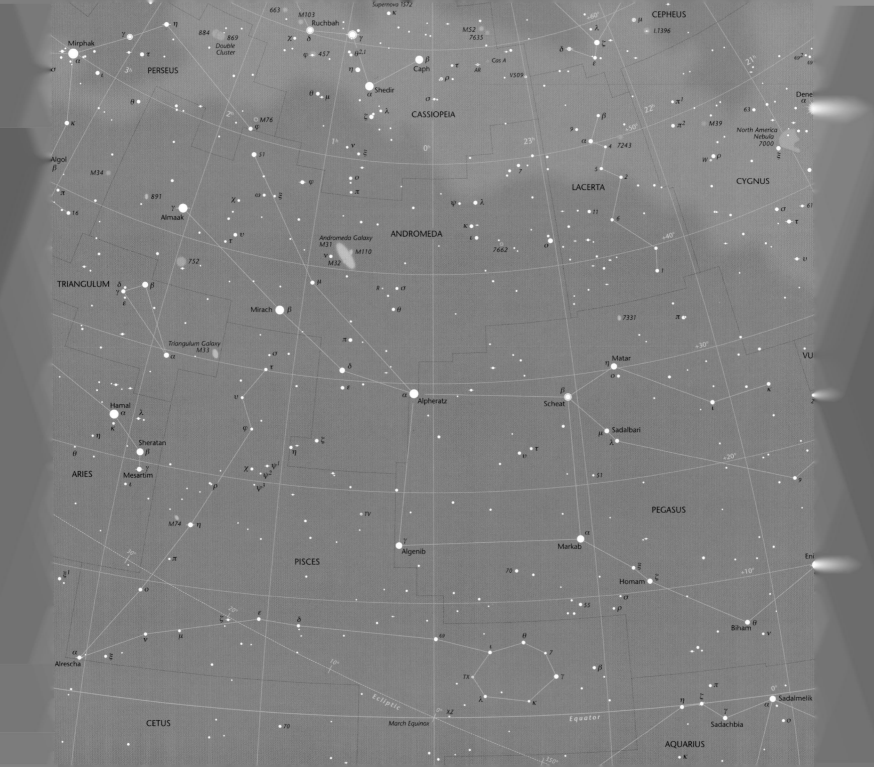

Chart 4

CAMELOPARDALIS

1528

M97
Owl Nebula M108 Merak
β

Muscida
ο

AX

24

15

7

δ

PERSEUS

TU

ξ

5ʰ

URSA MAJOR

11ʰ

CG

φ

2

+60°

+50°

ο

Capella α 1664

ε ξ

10ʰ

26 θ

15 9ʰ

27 8ʰ

21

ψ¹

ψ⁶

Menkalinan π β

ρ

η

λ

μ

ψ

λ Tania Borealis

μ

Tania Australis

κ ι

Talitha

LYNX

ψ⁹

ψ⁴

ψ⁵

ψ⁷ ψ²

2281 ψ³

+40°

ν τ

σ

M38
1907

φ

M36

β

31

10 UMa

UU

AURIGA

θ μ

χ

M37

21

10

38

2419

UU

WW

RT

+30°

θ

κ

β Elnath

LEO MINOR

α

σ⁴ σ²
σ³ σ¹

RS

τ

ο

π

α

Castor ρ

τ

κ

TAURUS

Rasalas

Adhafera

ξ

μ

ε

κ

55 ι

ρ² ρ¹

φ¹

φ² χ

χ

β σ

Pollux υ

ι

φ

GEMINI

ε Mebsuta

M35

90°

Crab Nebula M1

ξ

γ Algieba

λ Alterf

CANCER

ν

υ λ

ψ ω

κ

ω

100°

μ η Propus

ν

χ² U χ¹

+20°

η

LEO

2903

ξ₃

υ

μ

110

δ Wasat

ξ Mekbuda

2392
Eskimo Nebula

ν

2169

γ Asellus Borealis

η

120°

γ

Alhena

Regulus

ψ

Praesepe M44

δ θ

Asellus Australis

ζ

λ

ξ₃

α

ν

R

π

130°

χ

140°

Ecliptic

ο

30

S

+10°

Acubens

κ α M67

β

2264

μ

31

ο

π

ω

150°

Betelgeuse

α

SEXTANS

β

γ ε
Gomeisa β

η

13

2244
2237-39
Rosette Nebula

8

ORION

ω ζ ε
ρ δ

σ

Procyon α CANIS MINOR

δ³ δ²

18

θ

η

δ¹

2302

α

ι

τ²

τ¹

HYDRA

ζ

ξ

MONOCEROS

Equator

2232

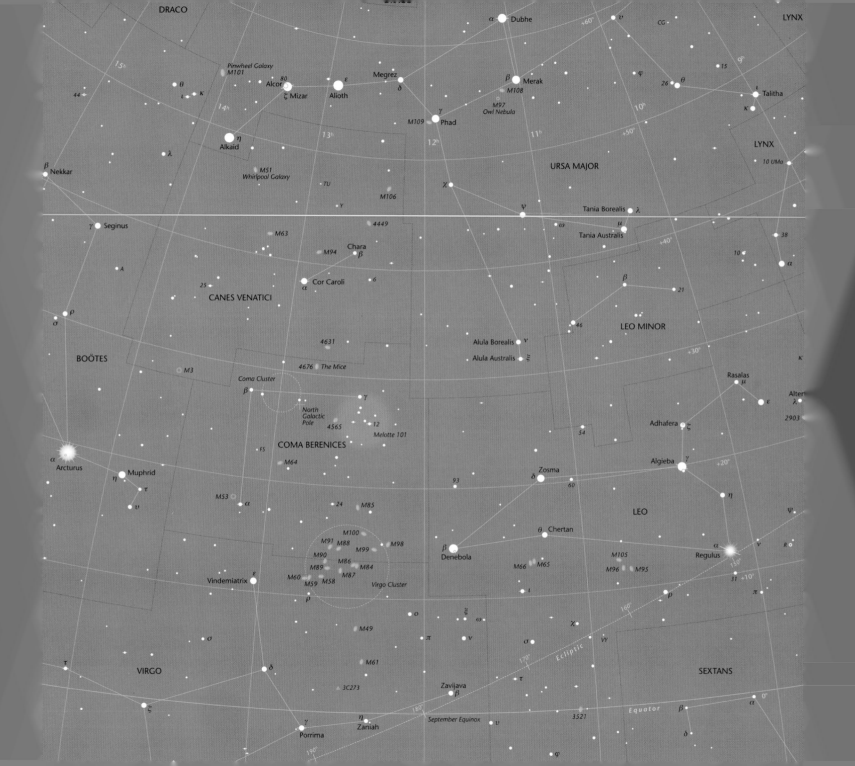

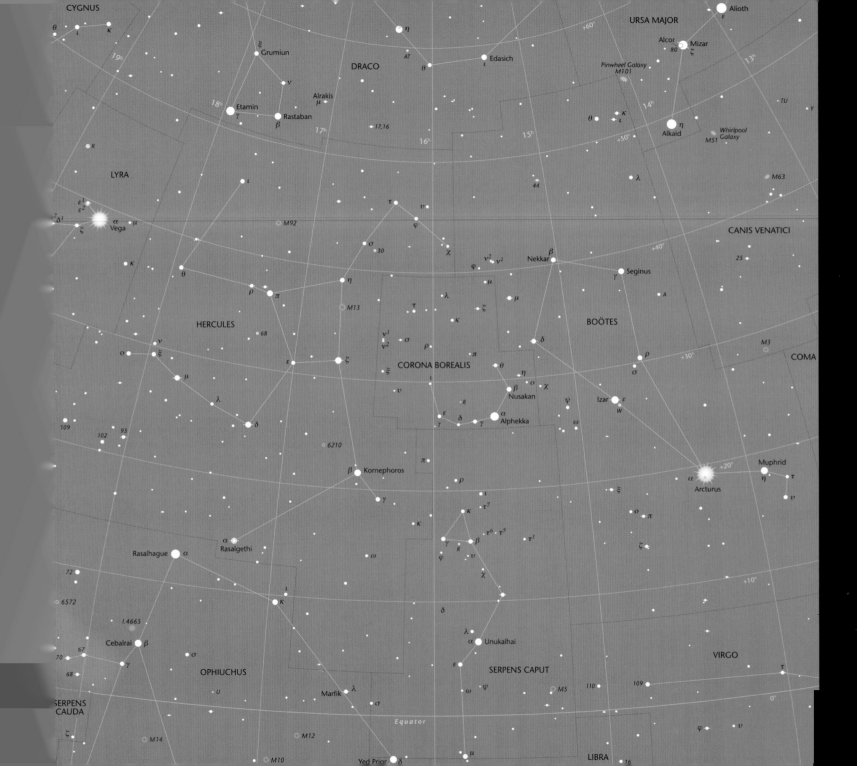

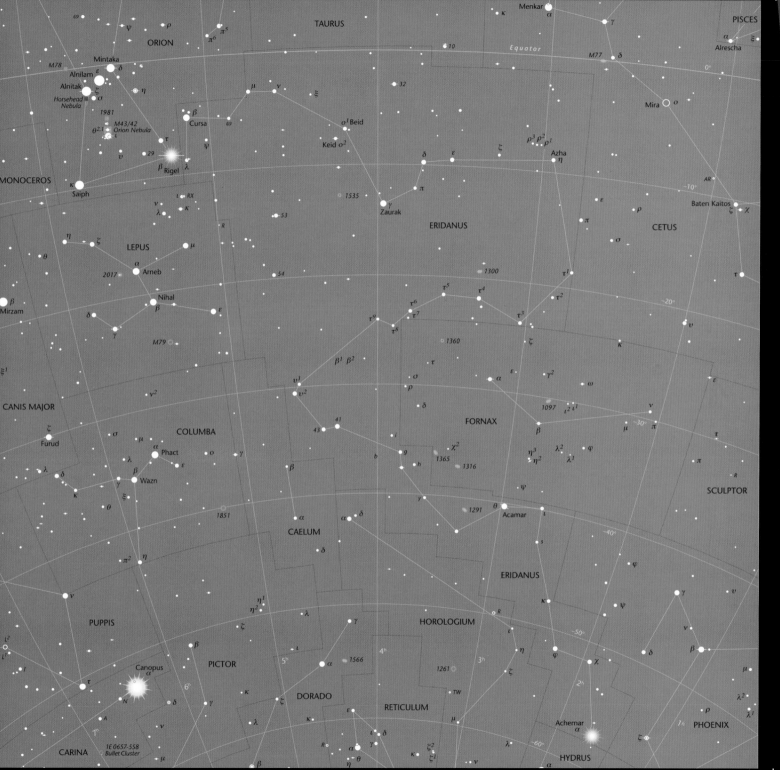

Chart 9

TAURUS

PISCES

ORION

Menkar
α
κ
γ
α
Alrescha
ξ

10
Equator
M77
δ
0°

Mintaka
M78
Alnilam ε δ
Alnitak ζ
Horsehead σ η
Nebula
1981
θ²·¹ M43/42
Orion Nebula
τ

β
Cursa
ω
o¹ Beid
Keid o²
ρ³ ρ²
ρ¹
Azha
Mira o

MONOCEROS
κ
Saiph
ι RX
ν λ κ
29
β Rigel λ
υ
53
1535
δ ε ζ
π
η
Zaurak
ERIDANUS
-10°
AR
ε
ρ
Baten Kaitos
ζ χ

CETUS

LEPUS
η ζ
θ
β
Mirzam
2017 α Arneb
Nihal β
γ
M79
ξ¹
54
μ ε
1300
τ⁵ τ⁴
τ⁶ τ⁷
τ⁹ τ³
τ⁸ ζ
1360 τ
β¹ β²
σ
τ¹
τ²
κ
υ
-20°
τ

CANIS MAJOR
ν²
υ¹
υ²
41
43
COLUMBA
ζ
Furud
σ μ
λ α Phact
β Wazn
κ ξ
θ
1851
π² η
ν
β
α
α δ
CAELEM
δ
b
g h
χ²
1365 1316
ρ
δ ε γ²
ω
1097 t² ι¹
β
η³ η²
ψ
γ
1291 θ
Acamar ι
s
FORNAX
ε
α
λ² φ
λ¹
ν
μ -30°
π
τ
π
R
SCULPTOR
ε

ERIDANUS
κ
φ
ψ
γ
ν
β
μ
λ²
λ¹

PUPPIS
ι² ι¹
ι
Canopus α
τ
N
δ
A
τ
1E 0657-558
Bullet Cluster
CARINA
PICTOR
η¹ η²
ζ
ι
5ʰ
γ
λ
1566
α
κ
ξ
DORADO
ν
λ
κ
4ʰ
1261
TW
3ʰ
R
ζ²
ι δ
ζ
μ
ν
θ
ζ
HOROLOGIUM
R
ι
η φ
ζ
2ʰ
Achernar α
ρ
λ²
λ¹
τ
PHOENIX
HYDRUS

RETICULUM
-40°
-50°
-60°

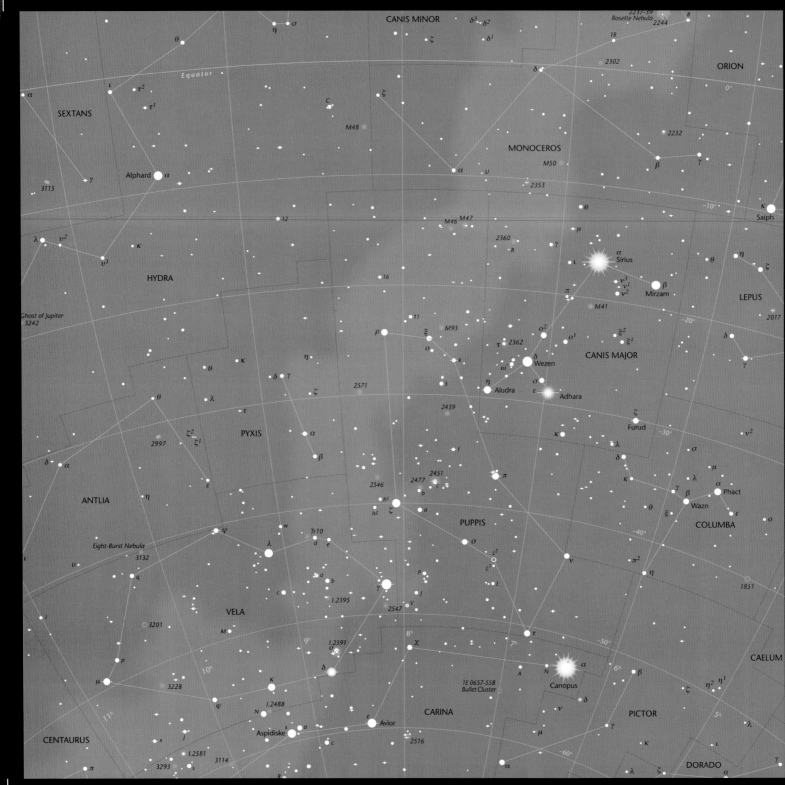

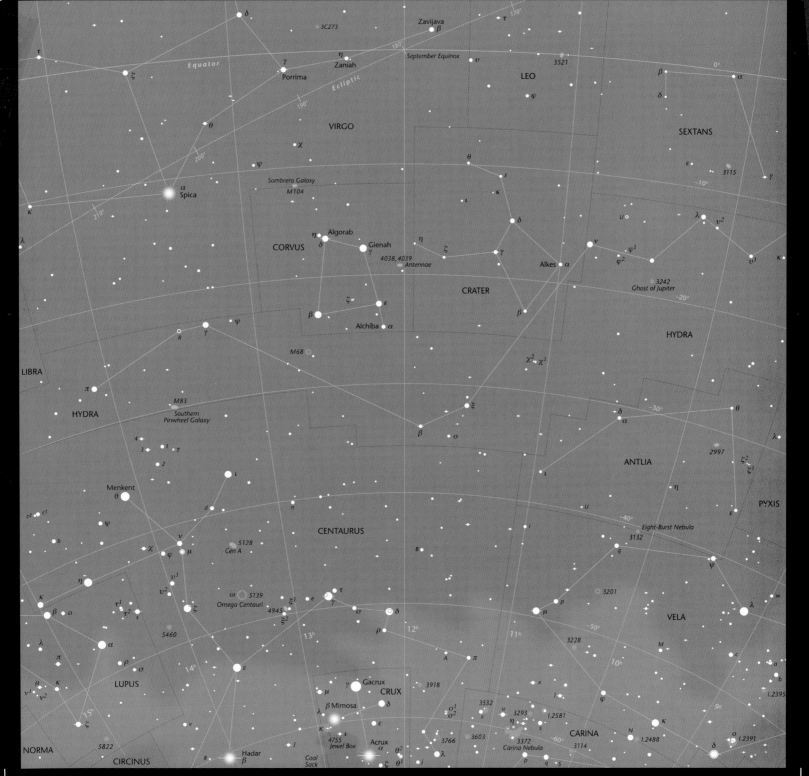

Chart 11

δ

3C273

Zavijava
β τ 170°

τ

Equator
γ
Zaniah
η
September Equinox
υ
3521
β 0°
α

Porrima
Ecliptic
LEO
SEXTANS

190°
φ
δ

θ
VIRGO
200°
χ
θ
ε
ε
3115
γ

ψ
ι
κ
−10°

Sombrero Galaxy
M104
U
λ υ²

α
Spica
δ
Algorab
η
ν
φ¹
υ¹ κ

210°
η
δ
Gienah
γ
CORVUS
4038,4039
Antennae
ξ
γ
Alkes α
φ²
3242
Ghost of Jupiter
−20°

κ
ξ
ε
CRATER
β

λ
β
Alchiba α
HYDRA

LIBRA
R
γ
ψ
M68
χ² χ¹

π
ξ
δ
−30°
θ

HYDRA
M83
Southern
Pinwheel Galaxy
β
ο
α
λ

4
3
1 T
ι
ANTLIA
2997
ξ²
ξ¹

2
ι
η

Menkent
θ
d
π
U
PYXIS

c² c¹
ν
CENTAURUS
i
−40°
Eight-Burst Nebula
3132
ε

ψ
χ φ μ
5128
Cen A
B
q

η
υ²
υ¹
ω 5139
Omega Centauri
e
τ
γ
3201

κ
β ο
τ¹
ξ
4945
ξ¹
ξ²
σ
δ
μ
ρ
VELA
M

τ² ι
5460
13ʰ
ρ
12ʰ
11ʰ
−50°
3228
c

λ
α
A
π
10ʰ

π
ρ σ
14ʰ
ε
γ Gacrux
3918
x
φ
b

μ
κ
μ Mimosa
δ
3532
u 3293
I.2581
N
κ
σ I.2391

ν¹ ν²
15ʰ
λ β
κ
ε
οᵁ¹
οᵁ²
x
η
3372
Carina Nebula
3114
I.2488
δ

NORMA
5822
CIRCINUS
Hadar
β
4755
Jewel Box
Coal
Sack
Acrux
α
θ²
θ¹
3766
3603
CARINA

R
ξ
j
ρ
q ς

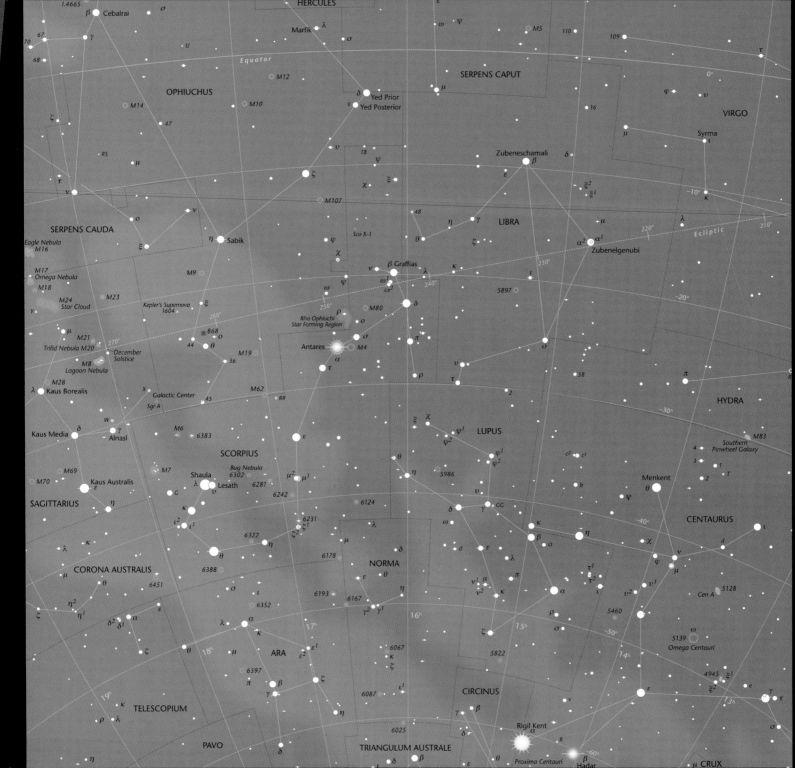

Chart 12

1.4665

β Cebalrai σ

67 γ

70

68 σ

λ

Marfik

ω ψ M5

110 109

τ

Equator

U σ

HERCULES

SERPENS CAPUT

φ υ

16 VIRGO

μ

OPHIUCHUS

M12

M14

δ Yed Prior

ε Yed Posterior

μ

μ Syrma

ι

M10

47

δ

Zubeneschamali

β

ζ

ν

18

ψ

−10°

κ

τ

ξ

ξ

ς²

ν

χ

M107

48

η γ

LIBRA

λ

σ

Ecliptic

RS

μ

ο

η

−220°

210°

SERPENS CAUDA

Sabik

ν

Sco X-1

θ

μ

α α¹

−210°

Eagle Nebula

M16

η

χ

ξ

ζ

Zubenelgenubi

−230°

M17

Ornega Nebula

M18

M9

ν

β Graffias

κ

ι

240°

−20°

M24

Star Cloud

ω

ω¹

λ

5897

M23

ξ

ω

ψ

ω²

γ

Kepler's Supernova

1604

250°

ρ

M80

δ

μ

M21

260°

ο

σ

π

σ

Trifid Nebula M20

B68

44

θ

ο

Rho Ophiuchi

Star Forming Region

σ

HYDRA

December

Solstice

36

M19

Antares

α

M4

π

−30°

M8

Lagoon Nebula

X

Galactic Center

M62

RR

τ

58

π

M28

45

Sgr A

2

λ Kaus Borealis

ρ

χ

M83

Kaus Media

δ

γ

M6

6383

ξ

ψ¹

LUPUS

Southern

Pinwheel Galaxy

Alnasl

ε

θ

ψ²

c²

c¹

Menkent

SCORPIUS

η

5986

b

θ

ψ

M69

M7

Bug Nebula

6302

μ²

μ¹

υ

ψ¹

CENTAURUS

M70

ε

Kaus Australis

Shaula

λ

Lesath

6281

6242

6124

δ

γ

GG

κ

η

ι

SAGITTARIUS

η

G

υ

6231

λ

μ

ω

ε

κ

χ

ν

φ

μ

ι²

ι¹

6322

ξ

τ¹

η

δ

β

ο

τ¹

ι

υ²

υ¹

Cen A 5128

CORONA AUSTRALIS

θ

6178

NORMA

π

ρ

5460

ω

λ

κ

θ

6388

6451

σ

η

δ

η

ν¹

μ

ξ

5139

η²

ι

6193

6167

κ

ν²

κ

σ

Omega Centauri

δ²

α

ε

6352

γ²

γ¹

16ʰ

ζ

5822

14ʰ

δ¹

λ

17ʰ

6067

4945

ξ²

ξ

κ

θ

μ

ARA

ε²

ε¹

κ

ι¹

γ

τ

19ʰ

18ʰ

6397

6087

CIRCINUS

ν

ε¹

κ

π

β

ε

TELESCOPIUM

γ

ζ

Rigil Kent

R

ρ

λ

6025

β

PAVO

δ

TRIANGULUM AUSTRALE

θ

β

σ

Proxima Centauri

Hadar

−60°

μ CRUX

Chart 14

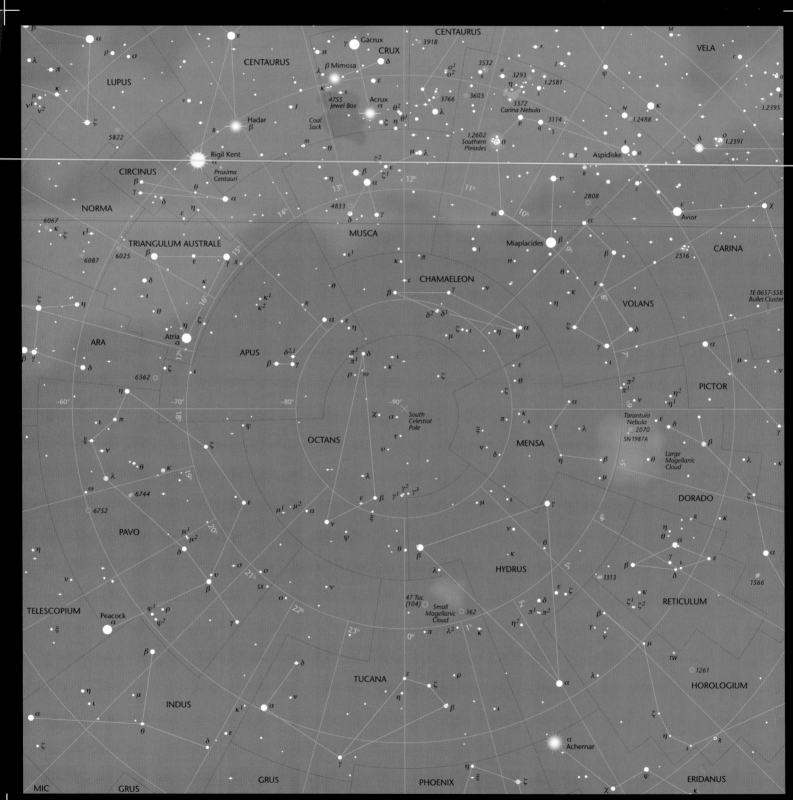

Name	Meaning	Abbreviation	Charts		Name	Meaning	Abbreviation	Charts
Andromeda	Andromeda	And	2		Lupus	Wolf	Lup	12
Antlia	Air Pump	Ant	11		Lynx	Lynx	Lyn	4
Apus	Bird of Paradise	Aps	14		Lyra	Lyre/Harp	Lyr	7
Aquarius	Water-bearer	Aqr	8, 13		Mensa	Table Mountain	Men	14
Aquila	Eagle	Aql	7, 13		Microscopium	Microscope	Mic	13
Ara	Altar	Ara	12,		Monoceros	Unicorn	Mon	4, 10
Aries	Ram	Ari	3		Musca	Fly	Mus	14
Auriga	Charioteer	Aur	3, 4		Norma	Carpenter's Square	Nor	12
Boötes	Herdsman	Boo	5, 6		Octans	Octant	Oct	14
Caelum	Chisel	Cae	9		Ophiuchus	Serpent-bearer	Oph	6, 7, 12
Camelopardalis	Giraffe	Cam	1, 3		Orion	Orion	Ori	3, 4, 9
Cancer	Crab	Cnc	4		Pavo	Peacock	Pav	14
Canes Venatici	Hunting Dogs	CVn	5		Pegasus	Pegasus	Peg	7
Canis Major	Greater Dog	Cma	10		Perseus	Perseus	Per	3
Canis Minor	Lesser Dog	Cmi	4		Phoenix	Phoenix	Phe	8, 9
Capricornus	Sea Goat	Cap	13		Pictor	Easel	Pic	9, 10
Carina	Keel	Car	10,		Pisces	Fishes	Psc	2, 8
Cassiopeia	Cassiopeia	Cas	1, 2		Piscis Austrinus	Southern Fish	PsA	8
Centaurus	Centaur	Cen	11, 12		Puppis	Poop Deck	Pup	10
Cepheus	Cepheus	Cep	1		Pyxis	Mariner's Compass	Pyx	10
Cetus	Sea Monster	Cet	3, 8, 9		Reticulum	Net	Ret	9, 14
Chamaeleon	Chameleon	Cha	14		Sagitta	Arrow	Sge	7
Circinus	Compass (for drawing)	Cir	14		Sagittarius	Archer	Sgr	13
Columba	Dove	Col	9		Scorpius	Scorpion	Sco	12
Coma Berenices	Berenice's Hair	Com	5		Sculptor	Sculptor	Scl	8
Corona Australis	Southern Crown	CrA	13		Scutum	Shield	Sct	13
Corona Borealis	Northern Crown	CrB	6		Serpens	Snake	Ser	6, 7, 12, 13
Corvus	Crow/Raven	Crv	11		Sextans	Sextant	Sex	4, 5, 10, 11
Crater	Cup	Crt	11		Taurus	Bull	Tau	3
Crux	Southern Cross	Cru	11, 14		Telescopium	Telescope	Tel	13
Cygnus	Swan	Cyg	7		Triangulum	Triangle	Tri	3
Delphinus	Dolphin	Del	7		Triangulum Australe	Southern Triangle	TrA	14
Dorado	Swordfish	Dor	9,		Tucana	Toucan	Tuc	14
Draco	Dragon	Dra	1, 6, 7		Ursa Major	Great Bear	Uma	1, 4, 5
Equuleus	Foal	Equ	7		Ursa Minor	Little Bear	Umi	1
Eridanus	Eridanus	Eri	9		Vela	Sails	Vel	10, 11
Fornax	Furnace	For	9		Virgo	Virgin	Vir	5, 6, 11, 12
Gemini	Twins	Gem	4		Volans	Flying Fish	Vol	14
Grus	Crane	Gru	8		Vulpecula	Fox	Vul	7
Hercules	Hercules	Her	6, 7					
Horologium	Pendulum Clock	Hor	9,					
Hydra	Sea Serpent	Hya	4, 10, 11, 12					
Hydrus	[Lesser] Water Snake	Hyi	14					
Indus	Indian	Ind	13,					
Lacerta	Lizard	Lac	2					
Leo	Lion	Leo	4, 5, 11					
Leo Minor	Lesser Lion	Lmi	5					
Lepus	Hare	Lep	9					
Libra	Scales	Lib	12					

The Greek Alphabet

α	Alpha	ι	Iota	ρ	Rho
β	Beta	κ	Kappa	σ	Sigma
γ	Gamma	λ	Lambda	τ	Tau
δ	Delta	μ	Mu	υ	Upsilon
ε	Epsilon	ν	Nu	φ	Phi
ζ	Zeta	ξ	Xi	χ	Chi
η	Eta	ο	Omicron	ψ	Psi
ϑ	Theta	π	Pi	ω	Omega

Photograph and Illustration Credits

Pg. 8, Dana Berry; 10 (UR), NASA/SDO/S. Wiessinger; 10 (LL), Royal Swedish Academy of Sciences/SST/Institute for Solar Physics; 10 (LR), NASA/SDO; 11, NASA/Goddard/SDO AIA Team; 12 (LL), NASA/JPL/Cornell; 12 (UR), NASA/JPL/Mosaic by Mattias Malmer; 13 (UL), NASA/Johns Hopkins University Applied Physics Laboratory/Carnegie Institution of Washington; 13 (ML), NASA/JPL/Malin Space Science Systems; 13 (LL), NASA Goddard Space Flight Center/Reto Stöckli; 13 (UR), NASA/Johns Hopkins University Applied Physics Laboratory/Carnegie Institution of Washington; 13 (LR), NASA/JPL-Caltech/ESA; 14, NASA/JPL/University of Arizona; 15 (UL), NASA/JPL; 15 (ML), NASA/JPL; 15 (LL), NASA/JPL/Space Science Institute; 15 (UR), NASA/JPL; 15 (LR), NASA/JPL-Caltech/Space Science Institute; 16 (UL), NASA/JPL/Space Science Institute; 16 (LL), NASA/JPL/DLR; 16 (UR), NASA/JPL/Space Science Institute; 17 (UL), NASA/GSFC/Arizona State University; 17 (LL), NASA/JPL/University of Arizona; 17 (UR), NASA/JPL-Caltech/University of Arizona/University of Idaho; 17 (LR), ESA/DLR/FU Berlin (G. Neukum); 18 (UL), W.M. Keck Observatory/Larry Sromovsky (University of Wisconsin); 18 (ML), NASA/JPL; 18 (LL/LR), NASA/JPL/Space Science Institute; 19 (UR), NASA/JPL/University of Arizona; 19 (ML/MR), NASA/JPL/Space Science Institute; 19 (LR), NASA/JPL/Space Science Institute; 20 (UL), NASA/JPL; 20 (UR), ESA/OSIRIS Team MPS/UPD/LAM/IAA/RSSD/INTA/UPM/DASP/IDA; 20 (ML), JAXA; 20 (MR), NASA/JPL/USGS; 20 (LL), NASA/JPL-Caltech/UMD; 20 (LR), NASA/JPL-Caltech/UMD; 21, NASA/JPL-Caltech/UCLA/MPS/DLR/IDA; 22 (UL), Jeffrey Pfau; 22 (I), Wikimedia Commons; 22 (LL), Wikimedia Commons; 22 (LR), Wikimedia Commons; 23 (UR), Getty Images; 23 (LR), Rob van Gent (University of Utrecht); 24 (LL-top), Wikimedia Commons; 24 (LL-bottom), Wikimedia Commons; 25, NASA; 26, NASA/ESA/E. Sabbi (STScI); 28 (I), ESO/APEX (MPIfR/ESO/OSO)/T. Stanke et al./Digitized Sky Survey 2; 28, W.H. Wang/IfA/University of Hawaii; 29 (UL/UR), NASA/JPL-Caltech/WISE Team; 29 (LR), ESO/Digitized Sky Survey 2/Davide De Martin; 30 (LL), Charles Messier/Mémoires de l'Académie Royale; 30 (UR), NASA/ESA/JPL-Caltech/IRAM; 30 (MR), ESO; 30 (LR), ESO/IDA/Danish 1.5 m/R.Gendler, J.-E. Ovaldsen, A. Hornstrup; 31, NASA/ESA/M. Robberto (Space Telescope Science Institute/ESA)/Hubble Space Telescope Orion Treasury Project Team; 32, ESO/VPHAS+ Consortium/Cambridge Astronomical Survey Unit; 33 (LL), NASA/ESA/N. Smith (University of California, Berkeley)/Hubble Heritage Team (STScI/AURA); 33 (LR), ESO/T. Preibisch; 34 (UL), NASA/ESA/Jeff Hester, Paul Scowen (Arizona State University); 34 (UR), NASA/ESA/Hubble Heritage Team (STScI/AURA); 35 (UR), ESO; 35 (LR), ESO; 36 (LL), T. A. Rector (University of Alaska Anchorage)/WIYN/NOAO/AURA/NSF; 36-37, ESA/PACS & SPIRE Consortium/Frédérique Motte/Laboratoire AIM Paris-Saclay/CEA/IRFU/CNRS/INSU/Université Paris Diderot/HOBYS Key Programme Consortia; 37 (LR), ESO/M.-R. Cioni/VISTA Magellanic Cloud Survey/Cambridge Astronomical Survey Unit; 38, T.A. Rector (University of Alaska Anchorage)/WIYN; 38 (UR), NASA/ESA/A. Nota (ESA/STScI, STScI/AURA); 38 (MR), ESO; 38 (LR), ESO/U.G. Jørgensen; 39 (UL), ESO; 39 (LL), ESO; 39 (UR), ESO (RCW 108), ESO; 39 (UR), NASA/ESA/Hubble Heritage Team (STScI/AURA); 39 (LR), University of Colorado/University of Hawaii/NOAO/AURA/NSF; 40, ESO/J. Borissova; 41(LR), NASA/ESA/Wolfgang Brandner, Boyke Rochau (MPIA)/Andrea Stolte (University of Cologne; 41, NASA/ESA/Hubble Heritage (STScI/AURA); 42 (LL), Lascaux Cave; 42 (UR), D. Bachmann; 42 (LR), NASA/JPL-Caltech/J. Stauffer (SSC/Caltech); 43, NASA/ESA/AURA/Caltech; 44 (UL), NASA/ESA/R. Sahai (JPL); 44 (LL), ESO; 44 (UR), A. Blaauw/BAN 44 (LR), NASA/JPL-Caltech/UCLA; 45 (UR), ESO; 45 (LR), ESO; 46, T.A.Rector (NOAO/AURA/NSF)/Hubble Heritage Team (STScI/AURA/NASA); 47 (UL), ESO; 47 (LL), NASA/ESA/Hubble Heritage Team; 47 (UR), ESO/J. Emerson/VISTA/Cambridge Astronomical Survey Unit; 48 (UR), ESO; 48 (LL), NASA/ESA/P. Hartigan (Rice University); 49 (UL), David Aguilar (Harvard-Smithsonian Center for Astrophysics; 49 (LL), Dana Berry; 49 (Disks), NASA/ESA/L. Ricci (ESO); 49 (LR), C.R. O'Dell/Rice University/NASA/ESA; 50 (LL), ALMA (ESO/NAOJ/NRAO)/NASA/ESA; 50 (MR), NASA/ESA/Digitized Sky Survey 2/Davide De Martin (ESA/Hubble); 51, ESA/NASA/L. Calçada (ESO); 51 (LR), NASA/ESA; 52 (UL), Wikimedia Commons; 52 (Lipperheij), Wikimedia Commons; 52 (ML), Cambridge University; 52 (LL), Yerkes Observatory; 52 (LR), NRAO/AUI/Dave Finley; 53 (UL), Unknown; 53 (LR), SKA; 54 (UL), LBTO; 54 (LL), Wikimedia Commons; 54 (LR), TMT; 55 (UL), Mount Wilson Observatory; 55 (UR), Palomar Observatory; 55 (LL), ESO/Serge Brunier; 55 (LR), Swinburne Astronomy Productions/ESO; 56, Lior Taylor; 58 (UL), United States Department of Energy; 58 (LL), Dana Berry; 59, Dana Berry; 60 (UL), Gemini Observatory/Lynette Cook; 60 (UR), NASA/CXC/GSFC/M.Corcoran et al./STScI; 60 (LR), NASA/ESA/F. Paresce (INAF-IASF)/R. O'Connell (University of Virginia, Charlottesville)/Wide Field Camera 3 Science Oversight Committee; 61, NASA/ESA/H. Richer (University of British Columbia); 61 (LR), Dana Berry; 62 (LL), NASA/SOHO; 62 (LR), Dana Berry; 62-63, ISAS/NASA; 63 (UL), Rijksmuseum Amsterdam; 63 (LR/LL), Digitized Sky Survey; 64 (UL), C. Barbieri (Univ. of Padua)/NASA/ESA; 64 (UR), ESA/Hubble/NASA; 64 (LR), NASA; 65 (UL), ALMA (ESO/NAOJ/NRAO)/M. Kornmesser (ESO); 65 (UR), Michael Liu (University of Hawaii); 65 (LL), NASA/JPL-Caltech; 66 (UL), Digitized Sky Survey; 66 (LL), NASA/Casey Reed; 66 (LR), Lucasfilm; 67 (UR), NASA/JPL-Caltech; 67 (MR), Dana Berry; 67 (LR), J. Trauger (JPL)/NASA/ESA; 68 (UL), Keck Observatory/Casey Reed; 68 (LL), Wikimedia Commons/Caelum Observatory; 68 (LR), David A. Hardy; 69, NASA/Hubble Heritage Team (AURA/STScI)/ESA; 70, NASA/ESA; 71, NASA/ESA/Martin Kornmesser (ESA/Hubble); 71 (UR), NASA/CXC/Univ. of Wisconsin-Madison/S.Heinz et al./DSS/CSIRO/ATNF/ATCA; 71 (LL), NASA/CXC/M. Weiss; 72, Dana Berry; 73 (UL), ESA/ATG medialab/ESO/S. Brunier; 73 (LL), ESO/L. Calçada; 73 (LR), NASA/Kepler mission/Wendy Stenzel; 74, NASA/ESA/G. Bacon (STScI); 75 (UL), ESA/Alfred Vidal-Madjar (Institut d'Astrophysique de Paris, CNRS, France)/NASA; 75 (LR), NASA/JPL-Caltech/T. Pyle (SSC); 76 (LL), ESO/L. Calçada; 76 (UR), Harvard-Smithsonian Center for Astrophysics/David Aguilar; 76 (LR), NASA/Kepler Mission/Dana Berry; 77 (UL), NASA/Ames/SETI Institute/JPL-Caltech ; 77 (LL), ESO; 77 (LR), NASA/Ames/JPL-Caltech; 78 (UL), ESO/L. Calçada; 78 (LL), NASA/Ames/JPL-Caltech; 78 (LR), NASA/ESA/G. Bacon (STScI); 79 (UL), ESO/L. Calçada; 79 (HD 69830), ESO; 79 (UR), NASA/JPL-Caltech/T. Pyle (SSC); 79 (LL), NASA/ESA/G. Bacon (STScI); 79 (LR), NASA/JPL-Caltech/T. Pyle (SSC); 80 (LL), ESO/L. Calçada; 80-81, NASA/Tim Pyle; 81 (LL), NASA/Ames/JPL-Caltech; 81 (LR), NASA/JPL-Caltech; 82, ESO/L. Calçada; 84 (UL), Dana Berry; 84 (LL), Jeff Bryant; 84 (LR), Dana Berry; 85 (UR), ESO and P. Kervella; 85 (LL), NASA/JPL-Caltech/UCLA; 85 (LR), Xavier Haubois (Observatoire de Paris) et al.; 86-87, NASA/ESA/C. Robert O'Dell (Vanderbilt University); 87 (UR), D. López (IAC); 87 (LR), Wikimedia Commons/Rawastrodata; 88, NASA/CXC/SAO/ STScI; 89 (UR), NASA/HST/UIUC/Y.Chu et al.; 89 (LR), Nordic Optical Telescope/Romano Corradi (Isaac Newton Group of Telescopes); 90, ESO/VISTA/J. Emerson/Cambridge Astronomical Survey Unit; 90 (LL), NASA/ESA/C.R. O'Dell (Vanderbilt University)/M. Meixner/P. McCullough/G. Bacon (Space Telescope Science Institute); 91 (All), NASA/NOAO/ESA/Hubble Helix Nebula Team/M. Meixner (STScI)/T.A. Rector (NRAO); 92 (UR), NASA/ESA/Hubble Heritage Team (STScI/AURA); 92 (ML), NASA/ESA/Hubble Heritage Team (STScI/AURA); 92 (LL/IC 4406), ESO; 92 (LL/NGC 6369), NASA/ESA/Hubble Heritage Team (STScI/AURA); 92 (LL/NGC 6362), ESA/Hubble/NASA; 92-93 (NGC 5189), NASA/ESA/Hubble Heritage Team (STScI/AURA); 93 (IRAS 12419-5414), ESA/NASA; 93 (M27), ESO; 93 (MyCn 18), Raghvendra Sahai, John Trauger (JPL)/WFPC2 science team/NASA/ESA; 93 (NGC 2392), NASA/ESA/Andrew Fruchter (STScI)/ERO team (STScI/ST-ECF); 93 (NGC 2346), NASA/ESA/Hubble Heritage Team (STScI/AURA); 94 (UR), NASA/ESA/H. Bond (STScI)/M. Barstow (University of Leicester); 94 (MR), NASA/R. Ciardullo (PSU)/H. Bond (STScI); 94 (LR), ESA/NASA; 95 (UL), Princeton University; 95 (LR), NASA/ESA/Hubble Key Project Team/High-Z Supernova Search Team; 96, ESO; 97 (UL), ESO/L. Calçada; 97 (LL), ESA/Hubble/NASA; 97 (LR), NASA/ESA/R. Kirshner (Harvard-Smithsonian Center for Astrophysics); 98, NASA/JPL-Caltech/WISE Team; 98 (I), Wikimedia Commons; 99 (LL), Wikimedia Commons; 99 (LR), NASA/ESA/R. Sankrit, W. Blair (Johns Hopkins University); 100, NASA/ESA/Allison Loll/Jeff Hester (Arizona State University)/Davide De Martin (ESA/Hubble); 100 (LL), Wikimedia Commons; 101, NASA/CXC/J. Hester , A.Loll (ASU) /JPL-Caltech/R. Gehrz (Univ. Minn.); 102 (LL), Crawford Collection; 102 (LR), NASA/JPL-Caltech/E. Dwek/R. Arendt; 103 (UL), NASA/JPL-Caltech/O. Krause (Steward Observatory); 103 (LL), NASA/CXC/MIT/T. Delaney et al.; 104 (LL), NASA/Swift/Cruz deWilde; 104-105, ESA/ECF; 105 (UL), Urania; 105 (LL), NASA/F. Walter (State

Index

Deep Space